CROP RESOURCES

Academic Press Rapid Manuscript Reproduction

Proceedings of the 17th Annual Meeting
of the Society for Economic Botany:
"Crop Resources"

The University of Illinois, Urbana
June 13–17, 1976

The Society for Economic Botany is an association
of individuals interested in the past, present,
and future uses of plants for man.

CROP RESOURCES

Edited by

David S. Seigler

Department of Botany
The University of Illinois
Urbana, Illinois

ACADEMIC PRESS, INC.
New York   San Francisco   London   1977
A Subsidiary of Harcourt Brace Jovanovich, Publishers

ACADEMIC PRESS, INC.
111 Fifth Avenue, New York, New York 10003

United Kingdom Edition published by
ACADEMIC PRESS, INC. (LONDON) LTD.
24/28 Oval Road. London NW1

Library of Congress Cataloging in Publication Data

Society for Economic Botany.
    Crop resources.

    1.   Field crops—Congresses.   I.   Seigler,
David S.   II.   Title.
SB183.2.S63   1977   338.1'7'3   77-13560
ISBN: 0-12-634950-9

PRINTED IN THE UNITED STATES OF AMERICA

CONTENTS

LIST OF CONTRIBUTORS

Numbers in parentheses indicate the pages on which authors' contributions begin.

Norman Applezweig (149), Norman Applezweig Associates, 442 W. 44th Street, New York, New York 10036

J. M. J. de Wet (111, 179) Department of Agronomy, The University of Illinois, Urbana, Illinois 61801

Norman R. Farnsworth (61), Department of Pharmacognosy and Pharmacology, College of Pharmacy, 830 Wood Street, The University of Illinois at the Medical Center, Chicago, Illinois 60612

Jack R. Harlan (105), Department of Agronomy, The University of Illinois, Urbana, Illinois 61801

T. Hymowitz (197), Department of Agronomy, The University of Illinois, Urbana, Illinois 61801

Ernest P. Imle (119), International Programs Division, Agricultural Research Service, U.S. Department of Agriculture, Hyattsville, Maryland 20782

Arnold Krochmal (75), Southeastern Forest Experiment Station, U.S. Department of Agriculture, Box 2570, Asheville, North Carolina 28802

Connie Krochmal (75), Southeastern Forest Experiment Station, U.S. Department of Agriculture, Box 2570, Asheville, North Carolina 28802

Robert L. Metcalf (165), Department of Entomology, The University of Illinois, Urbana, Illinois 61801

C. A. Newell (197), Department of Agronomy, The University of Illinois, Urbana, Illinois 61801

F. H. Otey (47), Northern Regional Research Center, Agricultural Research Service, U.S. Department of Agriculture, Peoria, Illinois 61604

L. H. Princen (1), Northern Regional Research Center, Agricultural Research Service, U.S. Department of Agriculture, Peoria, Illinois 61604

E. H. Pryde (25), Northern Regional Research Center, Agricultural Research Service, U.S. Department of Agriculture, Peoria, Illinois 61604

A. M. Rhodes (137), Department of Horticulture, The University of Illinois, Urbana, Illinois 61801

E. R. Sears (193), Agricultural Research Service, U. S. Department of Agriculture and Department of Agronomy, The University of Missouri, Columbia, Missouri 65201

Y. Shechter (179), Department of Biological Sciences, Lehman College, Bronx, New York 10468

C. Earle Smith, Jr. (79), Department of Anthropology, The University of Alabama, Box 6135, University, Alabama 35486

G. F. Sprague (97), Department of Agronomy, The University of Illinois, Urbana, Illinois 61801

George A. White (17), Germplasm Resources Laboratory, Agricultural Research Center, U.S. Department of Agriculture, Beltsville, Maryland 20705

Garrison Wilkes (211), Biology II Department, The University of Massachusetts, Boston Harbor Campus, Boston, Massachusetts 02215

PREFACE

It has recently become evident that many vital resources are being utilized at a greater rate than they are becoming available. Food resources are produced at the greatest rate ever, but population pressures have resulted in an overall decline in availability in many parts of the world. Other resources, such as petroleum, are nonrenewable and are rapidly becoming more expensive. In this volume, we have attempted to evaluate (a) the possible nonfood uses of cultivated plants, (b) the extent to which new and additional food resources may become available, (c) the prospects of several specialized uses of plants such as drugs, insecticides, rubber, and condiments, and (d) the origin of four major crops of the American midwest and prospects for their future development.

These papers were originally presented as a symposium on Crop Resources at the 17th Annual Meeting of the Society for Economic Botany in Urbana, Illinois, June 13–17, 1976. Most of them have been subsequently revised.

It is clear that we must turn to renewable resources to supplant a number of nonrenewable ones that are being depleted at a rapid pace. If we are to grow these new materials, however, several questions are raised. Do we have sufficient land on which to grow them? Can we afford to utilize land to grow materials for industrial use when we have an already insufficient food supply? Can new plants be domesticated? Can our currently available crops be selected and bred for even greater productivity? Can we expect the Green Revolution to rescue us from our difficult situation?

A number of crops currently used for a variety of purposes are in danger of being restricted by disease or, in some cases, by government regulations. Coffee, cocoa, rubber, *Dioscorea,* horseradish, and many others fall in this category. Can breeding programs select plants with sufficient disease resistance? Can these plants be supplanted by other materials that are equally satisfactory?

Agriculture in many areas of the world has become a monocultural system (or very nearly so). Furthermore, mechanization of planting, maintenance, and harvesting has led to the selection and development of crops with great uniformity and hence limited genetic diversity. In contrast, natural systems generally tend to favor diversity. Have we boxed ourselves in with regard to lack of diversity with our major crops? Do adequate gene resources exist to allow the development of new lines that are disease resistant, drought tolerant, etc.? In an effort to evaluate these possibilities, it is desirable to know more about the origin and domestication of crops in general and especially about our major cultivars today. Can we still utilize the original processes of domestication? Is there sufficient genetic diversity that is normally not expressed to permit selection of new lines? What indeed does the future hold for our major crops and hence our major food resources?

L. H. Princen has reviewed the possibilities of developing new crops from the view of a chemist. Many of these could supplant nonrenewable resources or those which must be imported into the U.S. In the following chapter, George White discusses the development of many of the same plants, but from the viewpoint of an agronomist. He evaluates what the possibilities of cultivation and acceptance may be. Many of these plants are potential sources of oils, lubricants, fibers for paper and other uses, rubber, gums, insecticides, antibiotics, and edible crop plants.

E. H. Pryde evaluates the use of currently cultivated oil-seed crops for industrial purposes. These are particularly important, since the long chains of the fatty acids of triglycerides of these oils are particularly desirable as replacements for the long-chain alkyl groups of petroleum. A substantial portion of our oil-seed production is currently used for industrial purposes such as coatings, plasticizers, nylons, lubricants, and glycerol. F. H. Otey discusses the industrial uses of carbohydrates, principally starch and cellulose, from a similar viewpoint. These materials are prime candidates as sources for the manufacture of a number of basic organic chemicals. We currently use about three-fourths as much starch and cellulose as we do petroleum. Therefore we must develop new sources if we wish to use these two materials for renewable resources. He optimistically concludes that we do have the land resources available to cultivate these crops without decreasing the availability of food crops.

Arnold and Connie Krochmal discuss their work in the development of specialized "cash" crops in the Appalachian area of the U.S. Most of these are used as medicinals or as food crops with primarily local uses.

Norman Farnsworth reviews the uses of plant materials as medicines. He concludes that this use has not declined as many are prone to think, but still represents a very major proportion of drug materials used today. In a related area, Norman Applezweig reviews the development and use of *Dioscorea* as a precursor for corticosteroids and contraceptives. In this case, government manipulation has largely been responsible for the decline of the crop. The future prospects of other plants as alternate sources of steroids look bright, however.

G. F. Sprague and Jack R. Harlan discuss the successes and short-comings of the Green Revolution. Their basic conclusions are that with-out population control, technological support within the country con-cerned, and study and control of social problems, the Green Revolution cannot succeed. J. M. J. de Wet discusses, in the following chapter, the likelihood that yields can be increased significantly in crop plants.

Because rubber is a major industrial product and because it can be cultivated, much recent interest has centered around plants that pro-duce latex. Natural rubber is used today in quantities greater than at any time in history, but it is also threatened by disease. Ernest Imle discus-ses the breeding programs currently in progress and the possibilities of introducing new germ plasm to create disease-resistant lines.

Robert L. Metcalf discusses the uses of plant materials for insec-ticides. Many biodegradable insecticides are derived from plants that are now cultivated in various parts of the world. Synthetic substitutes, based on structures of compounds derived from plants, have recently become of increasing importance because of concern about the ac-cumulation of certain synthetic compounds in the environment.

A. M. Rhodes discusses the current status of our most piquant sub-ject, horseradish, and its susceptibility to disease. New programs have been initiated to produce resistant lines so that this crop will be with us in the future. It is primarily cultivated in Illinois and adjoining states.

C. Earle Smith has reviewed the origin of several New World crops and this chapter sets the stage for those of de Wet and Shechter, Hymowitz and Newell, Sears, and Wilkes who discuss the four major crops of the Midwest: sorghum, soybeans, wheat, and corn. As all are of great importance, the origin of each and its prospects for the future are discussed in detail.

This book should be of interest to anyone with a concern for our natural resources, both renewable and nonrenewable. It should be of particular interest to agronomists, horticulturalists, chemists, chemical engineers, botanists, biologists, pharmacognosists, and anthro-pologists. Few other works exist in which the various aspects of the problem are surveyed in a single volume and by such prominent people in the respective fields of interest.

I would especially like to thank Dr. Jack Harlan, co-organizer of the symposium, for his ideas and assistance. Doctors William Tallent, John Beaman, J. M. J. de Wet, Ted Hymowitz, Norman Farnsworth, Harry Fong, and many others contributed to the organization of the program and ultimately to the publication of this book. I would also like to thank the authors of the chapters for providing manuscripts and for many suggestions and technical assistance.

CROP RESOURCES

POTENTIAL WEALTH IN NEW CROPS: RESEARCH AND DEVELOPMENT

L. H. Princen

Northern Regional Research Center
Agricultural Research Service
U.S. Department of Agriculture
Peoria, Illinois 61604

I. INTRODUCTION

There are several good reasons why Agriculture should always be on the lookout for potentially new crops from the vast resources of the kingdom of wild plants. Of the approximately 250,000 to 300,000 known plant species, less than 100 are grown commercially in the United States for food, feed, or industry at economic impact levels above $1 million. Worldwide, less than 300 species are so employed in organized agriculture. It is most likely that other wild plants could become economically viable if given a chance through proper research, development, and perhaps initial financial stimulation (1).

Another good reason for trying to develop new crops is the increasing reliance in the modern world upon exhaustable resources, such as natural gas, petroleum, and coal for energy and chemical raw materials. More than 500 pounds of petroleum per year per person is needed in the United States for chemicals and plastics production (Table 1). Ever increasing prices of these resources have made agricultural raw materials already quite competitive for several products (2). In the future, we will have to depend even more upon renewable resources, regardless of desirable prices, because of depletion or other temporary nonavailability of the traditional, mined raw materials. Our increased dependence upon petroleum imports is presented in Table 2 (3). That latest statistical information indicates that domestic oil production has been reduced to 8 million barrels per day in 1976, and that we must rely upon oil imports for even more than the 42% indicated by the trend in Table 2.

Modern agriculture tends to become exceedingly monocultural, with all economic and ecological dangers thereof. Many

1

TABLE 1

Use of Petroleum-Based Chemicals in the United States

Class of Chemicals	Consumption Billion Pounds[a]
Surfactants	5
Elastomers	6
Plastics	22
Lubricants	25
Polymers (adhesives, thickeners, flocculating agents, coatings)	40

[a]Total >500 pounds per person in the United States (total oil use is 8,500 pounds per person, of which over 40% is imported).

TABLE 2

Past and Predicted U.S. Petroleum Balances in Million Barrels/Day

Year	Domestic		Imports	
	Supply	Demand	To Balance	Percent of Total
1965	9.0	11.5	2.5	21.7
1970	11.3	14.7	3.4	23.1
1975	11.1	18.4	7.3	39.7
1980	11.8	22.5	10.7	47.5
1985	11.2	26.0	14.8	56.9

major parts of our country are already mentally associated with a single crop (or perhaps with at most two crops), such as wheat (Plains states), corn and soybeans (Midwest), cotton and peanuts (South), tobacco (Carolinas, Kentucky). Other areas can be easily associated with such crops as citrus, vegetables, peaches, apples or range grasses. Even if such monoculture practices might be most successful on the heavily mechanized farms of the United States, neither land nor equipment are available in many other countries of the world for such mass-production agriculture. Various areas within the USA are not suited for extensive monoculture farming either. With the development of specialty, low-volume new crops for

those labor-intensive, small farm areas, the United States and
other "bread-basket" countries could continue large volume
crops production to feed the world.

Other reasons for new crop development include the desirability of having available good protein sources for livestock
feeding in areas where established protein crops such as soybeans do not grow well (4). The northern and southeastern
parts of the United States are good examples for such needs.
Increased double-cropping practices make room for the introduction of new crop technology. The United States also relies
quite heavily upon several imported agricultural raw materials,
such as vegetable oils or natural rubber (Table 3) (5), which
might become temporarily inaccessible, due to climatic calam-
ity or political upheaval.

Whereas in the United States, about 340×10^6 acres
(14.8% of total land area) are being farmed for crops (Table 4
up to 470×10^6 acres is available immediately if needed (6).

TABLE 3

Some Imports of Agricultural and Related Materials, FY 1976

Commodity	Millions of Pounds per Year
Rapeseed oil	12.1
Tung oil	35.2
Waxes (carnauba, bees, candelilla)	11.5
Castor oil	99.9
Palm kernel oil	157.0
Coconut oil	1,103.8
Palm oil	1,041.6
Natural rubber	1,630.5

New crops can be and are being developed for the arid Southwest and other areas, where traditional crops do not do well.
Some estimates give figures in the order of 600×10^6 acres
that could ultimately be employed for cropland, if such were
economically desirable or desperately needed for maximum food,
feed, and industrial production. Even in our agriculturally
advantageous areas more acreage could be made available. In
more crowded countries, agriculture has advanced to include
production along and between four-lane highways or at large
jet-age airports through grazing, hay production or cultivation for grains and vegetables. Even in the United States,
the latter practice is now employed, such as at the municipal

TABLE 4

U.S. Land Use for Annual Crops

Crop	Planted Acres
Wheat	70.0×10^6
Corn	66.9
Hay	61.9
Soybean	53.6
Sorghum	15.5
Oat	13.6
Corn silage	9.7
Cotton	9.1
All other	36.3
Total	336.6×10^6
	(= 14.8% of total land area)

airport at Sioux City, Iowa, where soybeans were produced in 1976 between runways and taxiways. Table 5 shows that acreage needed to produce some new crops at desirable levels would not add significantly to the total cultivated acreage. The production levels are chosen in such a way that the crop would replace essentially all of the imported commodity or fabricated material now used in the United States.

TABLE 5

Acreage Needed for a Few Selected New Crops at Desired Production Levels

Crop	Amount of Production	Yield/Acre	Acreage
Crambe	30 million lbs.	1,000 lbs.	30,000
Kenaf	1 million lbs.	8 tons	125,000
Jojoba	100 million lbs.	2,000 lbs.	50,000
Lesquerella	200 million lbs.	1,000 lbs.	200,000
Stokes' aster	300 million lbs.	1,000 lbs.	300,000
Guayule	1,000 million lbs.	1,000 lbs.	1,000,000

The figures and thoughts presented above serve to show that neither in the United States nor in a major part of the rest of the world is agriculture with its back to the wall to feed the world population, and that it will be quite possible

to produce in an ecologically responsible way food and feed as
well as industrial commodities at economically acceptable
prices. Less than 700,000 acres of good cropland will supply
most of our needs in industrial products; the rest can be produced on desert and other waste lands.

II. NEW CROPS RESEARCH

The Agricultural Research Service (ARS) of the U.S. Department of Agriculture has conducted an active program of new
crops research for almost two decades (7). Although initially
organized to alleviate overproduction of established crops in
the 1950-1970 era, the program is no less viable now for the
completely different reasons presented here in the introduction. Of the major classes of chemicals derived from plants,
some are traditionally plentiful, whereas others are often in
short supply (Table 6); so the ARS screening program for new
crops from wild plants was developed to produce the most desirable chemicals, namely industrial oils and fibers for paper
making (8,9). More recently, screening for alkaloids with antitumor and pesticidal properties has also been added (10 and
C. R. Smith,Jr., R. G. Powell and K. L. Mikolajczak,
Chemother. Rep., in press). The program so far has resulted
in chemical analysis of seed samples from 6355 species for
oils and fatty acids, and evaluation of 506 species for fibers
as pulp, paper and board raw materials, 626 species for anti-
tumor agents almost 300 species for protein and amino acids
and about 70 species for pesticidal activity.

Seed analysis typically includes determination of seed
weight, oil content, infrared and ultraviolet spectroscopic
inspection, and HBr test of the oil, plus a nitrogen determination and starch test of the seed meal (11). Promising or
unusual oils are then further analyzed for their fatty acids
acomposition. Fiber analysis includes a visual and botanical
evaluation on crop potential, maceration for fiber, a micro-
scopic study and chemical analysis. Only if the species
appears to be a promising candidate is a full pulping evaluation carried out (12). Of the 506 species examined, only 92
were satisfactory to include in such a kraft cook evaluation.
Antitumor and pesticide screening is carried out on seed extracts and their fractions in cooperation with many other laboratories throughout the county. When activity is recognized,
the material is further characterized to arrive at the chemical structure of the active ingredient. More than 60 new
fatty acids and 40 other new chemicals have been discovered in
the program. Some of the screening results are described here
in more detail.

TABLE 6

Degree of Need for Botanical Chemicals in the United States

Class of Chemical	Need
Starch	Plentiful almost always worldwide
Cellulose	Probably always plentiful for most chemical use
Fibers	Shortage for pulp and paper as well as for lumber
Gums	Depend almost entirely on import, could be replaced partially by microbial gums
Fats and oils	Plentiful for food and feed purposes, shortage in specialty industrial oils
Waxes	Shortage for all uses
Protein	Shortage, but use could be improved by decreasing meat diet; dubious if real shortage exists
Hydrocarbons	Natural rubber entirely imported; even petroleum-based rubber in short supply
Specialty chemicals (alkaloids, essential oils, tannins, etc.)	Always need for medicinal, pesticidal or other special use products

A.   Long-Chain Fatty Acids

Although long-chain fatty acids are undesirable for food and feed purposes (scare of high-erucic acid content in rapeseed oil), they are eminently suitable for industrial production of polymers, lubricants, and plasticizers. Some of the more promising candidates for new crops with seed oils of high long-chain acid content ate given in Table 7 (13-18). If these plants were cultivated, they could replace imported rapeseed oil or petroleum products.

B.   Hydroxy Fatty Acids

For the supply of these acids, we rely almost entirely upon the annual import of 140 million pounds of castor oil. Several promising candidates for a native replacement crop have been found as shown in Table 8 (19,20).

TABLE 7

Species with Long-Chain Fatty Acids in Seed Oil

Species	Component in Triglyceride Oil
Crambe abyssinica	$60 + C_{22}$
Lunaria annua	$40\% \ C_{22}, \ 20\% \ C_{24}$
Limnanthes alba	$95\% \ C_{22} + C_{20}$
Selenia grandis	$58\% \ C_{22}$
Leavenworthia alabamica	$50\% \ C_{22}$
Marshallia caespitosa	$44\% \ C_{22}$

TABLE 8

Species with Hydroxy and Keto Fatty Acids

Species	Component in Triglyceride Oil
Lesquerella gracilis	14-OH-C_{20} (70%)
Holarrhena antidysenterica(?)	9-OH-C_{18} (70%)
Cardamine impatiens	Dihydroxy C_{22} and C_{24} (23%)
Chamaepeuce afra	Trihydroxy C_{18} (35%)
Lesquerella densipila	12-OH-C_{18} diene (50%)
Dimorphotheca sinuata	9-OH-C_{18} conj. diene (67%)
Coriaria myrtifolia	13-OH-C_{18} conj. diene (65%)
Cuspedaria pterocarpa	Keto acids (25%)

C.   Epoxy Fatty Acids

These acids have become increasingly important in plastics and coatings production. At present, over 140 million pounds of soybean oil are epoxidized yearly at great cost. These same acids are produced biochemically in a relatively large variety of wild plants as a major part of their triglyceride seed oils (Table 9) (21-24).

D.   Conjugated Unsaturation

Fatty acids with conjugated unsaturation are often advantageous as chemical intermediates for industrial products. Imported tung oil is a major source of conjugation. Conjugat-

ed unsaturation is also obtained by alkali isomerization of
soybean and linseed fatty acids (25). New potential sources
of conjugated polyunsaturation are listed in Table 10.

TABLE 9

Potential Sources of Epoxy Fatty Acids

Species	Epoxy Acid Content, %
Vernonia anthelmintica	68-75
Euphorbia lagascae	60-70
Stokesia laevis	75
Cephalocroton pueschellii	67
Erlangea tomentosa	50
Alchornea cordifolia	50 (C_{20})
Schlectendalia luzulaefolia	45

TABLE 10

Sources of Conjugated Unsaturation

Species	Type of Unsaturation
Valeriana officinalis	40% 9,11,13
Calendula officinalis	55% 8,10,12
Centranthus macrosiphon	65% 9,11,13
Impatiens edgeworthii	60% 9,11,13,15
Dimorphotheca sinuata	60% 10,12 (+ hydroxy)
Coriaria myrtifolia	65% 9,11 (+ hydroxy)

E.   Fiber Screening

For development of sorely needed additional supplies for
pulp and paper raw materials (70% of newsprint alone is imported), one has available two quite different routes. One is
used to a minor extent already, namely that of using agricultural residues such as grain straws and sugarcane bagasse, for
fiber production (26-29). The other alternative is to develop
completely new fiber sources from wild plants. Our survey has
resulted in many excellent candidates, of which only a few are
listed in Table 11. The most promising candidate is kenaf
(*Hibiscus cannabinus*) (30-32).

TABLE 11

Potential Nonwood Pulp and Paper Sources

Dual Purpose Crops	Special Fiber Crops
Grain straws	*Hibiscus cannabinus* (Kenaf)
(wheat, sorghum, corn)	*Althea rosea* (Hollyhock)
Sugarcane bagasse	*Sesbania exaltata* (Hemp Sesbania)
Flax straw	*Crotalaria juncea* (Sunn Hemp)

F.   Antitumor and Pesticide Activities

Many active fractions of extracts from seed and other plant components have been discovered. Several of the most promising leads have been carried through the various stages of development, including fractionation, isolation, chemical characterization, laboratory synthesis, and testing of additional derivatives. Testing of physiological activity is carried out typically against murine leukemia and solid tumors, selected fungi and other microorganisms, chicken louse, confused flour beetle, European corn borer, and malaria (33-36 and N. E. Delfel, H. B. Gillenwater, J. J. Ellis and D. Darrow, manuscript in preparation). Although the origin of active compounds is plant derived, production of the final products may not require any form of conventional agriculture. If a desirable compound with drug or pesticidal activity is discovered, it could conceivably be produced synthetically (37), or through tissue or cell culture (N. E. Delfel and J. A. Rothfus, manuscript in preparation).

III. CROP DEVELOPMENT

In the USDA new crops research program some experience has been gained in crop development of a few promising wild plants. Some of the experiences are summarized here to show crop potential and economic viability.

A.   *Crambe abyssinica*

This high-erucic acid-producing seed crop has been grown regularly for processing research purposes at levels of 1000 acres and more. Production has varied from 600 pounds per acre in Montana to as much as 2400 pounds per acre in Indiana and Illinois. In California, the yield has been as high as 4200 pounds per acre (38). Crambe is competitive with all traditional crops except corn at a sales price of 8 cents per pound (Tables 12 and 13). At that price and with a meal

credit of about $60/ton, the seed crushing plant should be
able to sell the oil competitively with respect to imported
high-erucic rapeseed oil. Production cost should compare with
all major crops except corn, and a net return of $40 to $60
should be realized. Various improvements can still be made in
several aspects of production, processing, and marketing.
Improved seed lines and more knowledge on preferred planting
dates and on herbicide and fertilizer applications are still
needed in the agronomical sector. Seed crushing and extraction parameters need to be investigated further for optimum
processing, especially in regard to enzyme destruction and
glucosinolate detoxification in the seed meal. Feeding
studies of the seed meal in cattle rations have to be complet-
ed for full FDA clearance (39). Also, a program for continuous availability of crambe products has to be devised.

TABLE 12

*Financial Comparison for Two New Crops
with Major Established Crops*

	Trend Yield	X	1975 Price	=	$ Gross Return/Acre
Corn	98.2	X	2.48	=	244
Sorghum	58.0	X	2.37	=	137
Soybeans	27.1	X	4.60	=	125
Barley	43.0	X	2.79	=	120
Wheat	32.0	X	3.54	=	113
Crambe	1,500	X	0.08	=	120
Kenaf	8	X	35.00	=	280

B.   *Hisbiscus cannabinus*

    Until 1976, kenaf had not been grown semicommercially;
however, on plots of several acres in size, it has produced
from 5 to 10 tons of dry harvested matter per acre (40). This
growth rate translates into twice the amount of fiber production per year per acre as is obtained from traditional tree
farming operations. Production cost should be about $100 per
acre, much like corn, and provide a healthy net return for the
farming operation (Tables 12 and 13). Studies on harvesting,
storing, transporting, compacting, pulping, and papermaking of
green and of frost or drought-killed kenaf (41) have shown
this plant species to be eminently suitable for replacing or
augmenting traditional fiber crops at competitive prices

(Table 11). The only major problem with kenaf is production
loss in areas infested with root knot nematodes, but several
resistant lines are being developed. Also, commercially
available nematocides could be used advantageously. In 1976,
500 acres of kenaf has been grown commercially for full-scale
evaluations. In 1977, the crop may be increased to 2000 acres.

TABLE 13

Production Cost and Net Return per Acre
for Established Crops

Direct Cost[a] per Acre, 1976		Net Return per Acre
Corn	137-153	91-107
Sorghums	84-97	40-53
Soybeans	65-80	45-60
Barley	63-69	51-57
Wheat	57-65	48-56

[a]Purchased production items, machinery ownership, and farm
overhead.

C.   *Limnanthes alba*

     Meadowfoam, a native of the northwestern Pacific states,
is another producer of seed oil with long-chain fatty acids
(C_{20} and C_{22}). One upright growing cultivar has been developed and registered at Oregon State University, and seed increase and small production attempts are underway. No production or economic figures are available at this time, but they
should be comparable to those of crambe. More work needs to
be done on natural toxicants in the seed meal, and on processing and product development of the oil. The business and
farming conditions are such in Oregon and Washington that meadowfoam production may increase rapidly in that area.

D.   *Simmondsia chinensis (or californica ?)*

     Jojoba has been known for a long time as the producer of
a unique seed oil. That oil consists of 100% liquid wax ester,
and it should be an excellent replacement for sperm whale oil
in lubricants and many other specialty uses. Because jojoba
is a slow-growing dioceous shrub of the arid southwest, crop
development is by necessity a long-term process. Interest in
jojoba is strong in many arid areas of the world, such as
Israel, Mexico, Australia and South Africa ("Proccedings of
the 2nd International Conference on Jojoba and its Uses" (T. K.

Miwa, Ed.), in press). Mexico appears farthest ahead in developing jojoba as a crop, with some 1700 acres under cultivation. The first harvestable crop from seeded plantations is expected in 1979. Developments in the United States are spearheaded by the Bureau of Indian Affairs, Department of the Interior, because jojoba could be an excellent cash crop for the southwest Indian tribes.

E. *Parthenium argentatum*

Guayule has been cultivated successfully in the Southwest during World War II for the production of natural rubber. After the war, the project was abandoned when hevea rubber from Southeast Asia again became available. With the increased consciousness that the United States is becoming more and more dependent upon foreign raw materials for its industrial production, attention has again been directed toward guayule as a source of natural rubber. Jojoba and guayule could be excellent companion crops for the arid Southwest, where there would be no competition with other crops. Other hydrocarbon producing plants are also being screened and considered for development (42).

F. Other New Crops

Other attempts of new crop development include *Stokesia*, *Lesquerella*, and *Crotalaria* species, but none of these have been grown in sufficient quantities or under enough diverse conditions to indicate if there are any particular unsolvable problems involved. In cultivation of wild plants, such problems as seed scattering, lack of flowering or seed setting, poor upright growth habit, or disease, often have to be overcome through breeding or selecting of cultivars and by extensive agronomic research.

IV. DISCUSSION AND CONCLUSION

The USDA screening program of wild plants for useful products for industrial, medicinal, pesticidal or food and feed purposes has so far unearthed many new chemical compounds and quite a number of potentially useful new crops. From the figures given here on the numbers of plant species screened for various compounds and activities, and from the increased chemical and botanical knowledge gained, it becomes at once obvious that the screening program has been successful as a research effort, and that it should be continued to make us completely familiar with the chemical riches of the plant kingdom.

Although only a few potential new crops have been brought through all stages of development, and although no new crops

have been completely accepted as yet, it appears that the concept itself is viable. At this point, the market place with
its economic and political forces will have to determine the
desirability of further development into full production for
any one of the described new crops. In this presentation it
has been shown that (1) wild plants can be selected and developed into useful, desirable, and well-producing new crops; (2)
such development requires approximately 15 years of cooperative work by many scientists and agencies; (3) the acreage
needed for new crops is available today and does not have to
reduce our food production; and (4) the availability of new
crops for industry does not necessarily result in immediate
adoption.

The latter point may require some further thought. A new
crop could be compared with a newly developed industrial product. Such a product requires a certain amount of financial
commitment on the company's part to bring it from the development stage into production and marketing. Such a commitment
is traditionally made from corporate funding other than the
research budget. Perhaps, a similar commitment will be required, from either the private or the public sector, to provide new crop support through the initial transition period.
The length of such a period should vary with the kind of crop
under consideration, but should typically be in the order of
5 years. Thereafter, the market forces could be allowed to
take over completely, when farmer, processor, and consumer
have become familiar with all aspects of the new crop and its
ultimate products.

V.   REFERENCES

1. Wolff, J. A. and Jones, Q. *Chemurgic Dig.* 17, 4 (1958).
2. Pryde, E. H., This volume.
3. Dupree, W. G., Jr., and West, J. A. "U.S. Energy Through
    the Year 2000," U.S. Department of the Interior, Washing-
    ton, D.C. 1972.
4. VanEtten, C. H., Kwolek, W. F., Peters, J. E., and
    Barclay, A. S. *J. Agric. Food Chem.* 15, 1077 (1967).
5. U.S. Foreign Agricultural Trade Statistical Report,
    Fiscal Year 1976, Economic Research Service, U.S. Depart-
    ment of Agriculture, Washington, D.C., 1976.
6. Agricultural Statistics. p. 420, U.S. Department of Agri-
    culture, Washington, D.C., 1975.
7. Wolff, I. A. *Econ. Bot.* 20, 2 (1966).
8. Earle, F. R., Melvin, E. H., Mason, L. H., VanEtten, C.H.,
    Wolff, I. A., and Jones, Q. *J. Am. Oil Chem. Soc.* 36,
    304 (1959).
9. Nieschlag, H. J., Nelson, G. H., Wolff, I. A., and Perdue,
    R. E., Jr. *Tappi* 43(3), 193 (1960).

10. Hagemann, J. W., Pearl, M. B., Higgins, J. J., Delfel, N. E., and Earle, F. R. *J. Agric. Food Chem.* 20, 906 (1972).

11. Barclay, A. S. and Earle, F. R. *Econ. Bot.* 28, 178 (1974).

12. Cunningham, R. L., Clark, T. F., Kwolek, W. F., Wolff, I. A., and Jones, Q. *Tappi* 53(9), 1697 (1970).

13. Smith, C. R., Jr., Bagby, M. O., Miwa, T. K., Lohmar, R. L., and Wolff, I. A. *J. Org. Chem.* 25, 1770 (1960).

14. Wilson, T. L., Smith, C. R., Jr., and Wolff, I. A. *J. Am. Oil Chem. Soc.* 39, 104 (1962).

15. "Crambe--A Potential New Crop for Industrial and Feed Uses." ARS 34-42, Agricultural Research Service, U.S. Department of Agriculture, Washington, D.C., 1962.

16. Bruun, J. H. and Matchett, J. R. *J. Am. Oil Chem. Soc.* 40, 1 (1963).

17. Miller, R. W., Daxenbichler, M. E., Earle, F. R., and Gentry, H. S. *J. Am. Oil Chem. Soc.* 41, 167 (1964).

18. Phillips, B. E., Smith, C. R., Jr., and Tallent, W. H. *Lipids* 6,93 (1971).

19. Smith, C. R., Jr., Wilson, T. L., Miwa, T. K., Zobel, H., Lohmar, R. L., and Wolff, I. A. *J. Org. Chem.* 26, 2903 (1961).

20. Mikolajczak, K. L., Earle, F. R. and Wolff, I. A. *J. Am. Oil Chem. Soc.* 39, 78 (1962).

21. Smith, C. R., Jr., Koch, K. F., and Wolff, I. A. *J. Am. Oil. Chem. Soc.* 36, 219 (1959).

22. Phillips, B. E., Smith, C. R., Jr., and Hagemann, J. W. *Lipids* 4, 473 (1969).

23. Earle, F. R. *J. Am. Oil. Chem. Soc.* 47, 510 (1970).

24. White, G. A. and Earle, F. R. *Agron. J.* 63, 441 (1971).

25. DeJarlais, W. J., Gast, L. E., and Cowan, J. C. *J. Am. Oil Chem. Soc.* 50, 18 (1973).

26. Ernst, A. J., Nelson, G. H., and Knapp, S. B. *Tappi* 40, 873 (1957).

27. Ernst, A. J., Clark, T. F., Finkner, M. D., and Dutton, H. J. *Tappi* 42(3), 235 (1959).

28. Naffziger, T. R., Matussewski, R. S., Nelson, G. H., and Clark, T. F. *Tappi* 42(7), 609 (1959).

29. Ernst, A. J., Fouad, Y., and Clark, T. F. *Tappi* 43(1), 49 (1960).

30. Nieschlag, H. J. Nelson G. H., and Wolff, I. A. *Tappi* 44(7), 515 (1961).

31. Clark, T. F., Nelson, G. H., Nieschlag, H. J., and Wolff, I. A. *Tappi* 45(10), 780 (1962).

32. Clark, T. F. in "Pulp and Paper Manufacture. II. Control, Secondary Fiber, Structural Board Coating," 2nd Edition, 1969.

33. Powell, R. G., Weisleder, D., Smith, C. R., Jr., and

Wolff, I. A. *Tetrahedron Lett.* (46) 4081 (1969).

34. Mikolajczak, K. L., Powell, R. G., and Smith, C. R., Jr. *Tetrahedron* 28, 1995 (1972).

35. Powell, R. G., Weisleder, D., and Smith, C. R., Jr. *J. Pharm. Sci.* 61(8), 1227 (1972).

36. Powell, R. G., Smith, C. R., Jr., and Madrigal, R. V. *Planta Medica* 30, 1 (1976).

37. Mikolajczak, K. L., Smith, C. R., Jr., and Weisleder, D. Tetrahedron Lett. (3) 283 (1973).

38. White, G. E. and Wolff, I. A. *Crops Soils* 19(4), 16 (1967).

39. VanEtten, C. H., Daxenbichler, M. E., Schroeder, W., Princen, L. H. and Perry, T. W. *Can. J. Anim. Sci.*, in press, March 1977.

40. White, G. A., Cummins, D. G., Whitely, E. L., Fike, W. T., Greig, J. K., Martin, J. A., Killinger, G. B., Higgins, J. J., and Clark, T. F. Prod. Res. Rep. No. 113, Agricultural Research Service, U.S. Department of Agriculture, Washington, D.C.

41. Clark, T. F. and Bagby, M. O. Indian Pulp and Paper Technical Association, Conference Number Supplement, Vol. 7, 16, 1970.

42. Buchanan, R. A., Cull, I. M., Otey, F. H., and Russell, C. R., Submitted for publication.

PLANT INTRODUCTIONS--A SOURCE OF NEW CROPS

George A. White

Beltsville Agricultural Research Center
Agricultural Research Service
U.S. Department of Agriculture
Beltsville, Maryland 20705

I. INTRODUCTION

Most major crops of the United States are plant introductions. Among agronomic crops, sunflower is the only notable exception. Of plants native to the U.S., only sunflower, artichoke, and a few small fruits are of major importance as world crops. The economic importance of native grasses is difficult to assess.

Maize was moved considerable distance from Mexico or Central America by migrating Indians and was sparsely cultivated well before white settlers arrived on the East Coast. Accounts vary as to when and how rice was introduced. In one instance, a ship from Madagascar made an unplanned docking under adverse conditions, and befriended by settlers, the captain presented a handful of rice seed as a token of appreciation. This occurred in South Carolina in 1694. Rice became established at about that time. Other reports indicate that trial plantings of rice were made in Virginia as early as 1609. Now, rice is an important export crop. The Soybean, perhaps, should be called the "golden" crop of plant introduction. After introduction in about 1800, it became initially important as a hay crop but eventually rose from limited acreage for beans in 1900 to about 56 million harvested acres in 1973.

Relatively recently-established introduced new crops include guar, safflower, broccoli, and pistachio. The specialty crops, chinese waterchestnut and chinese gooseberry, are produced on a very limited scale. In the case of the gumseed crop, guar, No. 9666 from India in 1903, is the first P.I. (Plant Introduction) in our documentation system. In 1970, an estimated 90,5000 acres were harvested, although the potential appears

to be about 400,000 acres. The first American experiment station trial with safflower was conducted in California in 1899.
The first recorded P.I. of safflower was number 993 introduced
in 1898 from Turkestan. Safflower became commercialized in
about 1948, had a peak acreage of about 300,000 acres, and
about 220,000 acres were harvested in 1974.

A.   New Crop Development Program

In 1956, a systematic program was initiated to develop
new crops. Emphasis, because of crop surpluses at that time,
was placed on replacement crops, especially those with industrial use. Since 1956, many new crop leads have been identified. More than 225 species have undergone preliminary field
testing. Advanced developmental efforts, both agronomic and
utilization, have been extended to about 12 genera, encompass-
ing more than 60 species and botanical varieties (1).

While the emphasis on industrial-use crops remains today,
more efforts are being directed to medicinal--especially anti-
tumor compounds, drugs, and to natural products. These
efforts generally are special programs with outside funding.
Work on procurement, chemical testing, and propagation of several plants with anti-tumor properties was covered in a 1975
Economic Botany Symposium; hence, no further consideration
will be given here.

A current example of a special program to develop new
crops is the work on *Papaver bracteatum* Lindl. This attractive perennial of Iranian origin contains the chemical,
thebaine, in fruits, other aerial parts, and roots. Thebaine
is an alternative route to codeine that is less subject to
drug abuse than the opium poppy. Preliminary field results
show good potential with high crop value for limited acreage.

Interest in new crops sometimes may be on a state basis.
For instance, *Plantago ovata* Forsk., a source of seed mucilage
for laxatives and other products, has good potential to become
firmly established as a new crop in Arizona. Problems of disease, seed-shattering, and insufficient winter hardiness have
been partially solved.

1.   *Fiber Crops as sources of paper pulp*

Of many screened species, only a few were found to combine suitable chemical and agronomic characteristics to justify intensive developmental research. Kenaf (*Hibiscus
cannabinus* L.), a diploid species from Africa, is the leading
candidate. Its tetraploid relative, roselle (*H. sabdariffa* L.),
has received some attention because of a natural resistance to
root-knot nematode (2). Renewed interest in sunn hemp
(*Crotalaria juncea* L.), a legume from India, has resulted from
the increasing cost of nitrogen fertilizer. This species is

also resistant to nematodes. Certain sorghums have very high
yield (biomass) potential but more restricted usage than the
above species (3).

About 500 acres of kenaf were planted in New Jersey in
1976. Late planting, coupled with very dry conditions, will
likely restrict yields. About 1,000 acres are projected for
North Carolina in 1977. There are eight named cultivars of
kenaf that have been field-tested at several locations.
Dr. Charles Adamson, kenaf breeder from the USDA's Plant Introduction Station, Savannah, Georgia, has selected six superior
nematode-tolerant breeding lines. These will be increased in
1976-77 and field-tested at several locations in 1977.

Production and harvesting methods for kenaf will vary
according to the end usage, the area involved, and storage of
the raw material. While the whole stem can be pulped, there
is considerable emphasis on the use of only the best fiber
portion of the stem. Late planting in dense stands results in
improved weed control and ease of harvest. Small-diametered
stems should also give a relatively higher ratio of bast to
wood fiber. The crop is field-dried after a killing frost,
cut with a hay conditioner, baled, and decorticated. Variant
approaches would include chemical killing of plants (more
southerly locations) and window-drying of green cut plants.
When whole stalks are to be pulped, chopping of the green or
dry crop is one harvest method (4).

Cultural refinements to maximize yields are needed, especially with respect to weed control, date of planting, and
plant density. Seed production research is underway.

Several years ago, research on *Crotalaria juncea* was ter-
minated because of the better yield and breeding potential of
kenaf. This species does, however, produce a quality fiber
(some is used in pulp in India and, possibly, Brazil) and
field-dries very rapidly after being frost-killed. This would
be an important harvesting factor (5). Tall plants tend to
branch near the tops and may lodge badly. Germplasm of this
species is quite limited. Breeding efforts would be needed to
make this species a potentially viable crop.

The future of kenaf appears bright. Successful commer-
cialization depends on economic feasibility as reflected in
the value of the fiber and the crop yield.

2. *New Oilseed Crops*

Crambe abyssinica Hochst. ex R. E. Fries has been successfully grown on a field scale in Ohio, Indiana, Illinois, North
Dakota, Montana, Wyoming, Washington, Oregon, California, and
Texas. About 1,000 acres were harvested in 1975 and approximately the same number planted in 1976. Production from 100
to more than 1;500 acres has been sustained for several years.

Three cultivars, 'Prophet', 'Indy', and 'Meyer', have been released over a several year period by the Purdue University Agricultural Experiment Station. A few superior breeding lines in this program are promising for future release. Lack of program resources, especially breeding and germplasm, greatly hampered improvement of crambe after recognition of the excellent use-agronomic potential of crambe. Unfortunately, the distribution of wild populations of *C. abyssinica* remains rather obscure even today. We have only known of the occurrence of the species in Turkey within the last three years and have seed from only three populations. The section Leptocrambe, to which *C. abyssinica* belongs, contains six species. The four species available to us all contain high levels of erucic acid in their seed oils. *C. abyssinica* has the best crop potential, especially with respect to seed retention. The other species do offer breeding potential because of different ploidy levels and might contribute, among other characteristics, more cold tolerance (6,7).

TABLE 1

Composition of Crambe seed

Species	Range in composition (%)		
	Oil	Erucic acid	Hull
C. hispanica var. *hispanica*	23-41	55-64	28-45
C. hispanica var. *glabrata*	30-39	47-61	32-49
C. filiformis	30-38	54-60	40-63

These data provided by R. Kleiman, Research Leader, USDA, ARS, Northern Regional Research Center, Peoria, Illinois. Data are not yet available for all accessions.

In Table 1, results from chemical analysis of a portion of a 1974 collection of two species are shown. These values do not differ greatly from those of *C. abyssinica*. The hull percentages for *C. filiformis* Jacq. are high.

The future of *Crambe* as a crop is uncertain. While it has good agronomic characteristics, yield improvement, disease resistance, and cultural refinements for specific production areas are needed. Processing facilities should be close to the production area. The oil has excellent qualities; it can, and has, been used in commercial products. Uses would likely expand with dependable supplies. Efforts involving improvement through breeding, cultural refinements and the assemblage, evaluation, and preservation of *Crambe* germplasm should continue (6,7).

Meadowfoam is viewed commonly along meadow streams in northern California and southern Oregon. The spring show of meadowfoam consists primarily of *Limnanthes alba* Hartw. and its botanical varieties. There are nine Limnanthes species (and 11 varieties), all but one of which have been field-tested for agronomic traits. Members of the section Inflexae, which includes *L. alba*, have the best potential because of superior seed retention. Research, with emphasis on *L. alba*, is underway in Oregon, California, and Maryland. The Oregon Agricultural Experiment Station has registered the cultivar 'Foamore' and has some superior breeding lines under test. Cultural methods for this winter-annual crop have been worked out. Chemical weed control will be required. Increased combinable yield is the greatest crop development need. Intensive selection for improved seed retention will help. Breeding objectives also include higher seed oil content in agronomically superior lines (8,9).

Lesquerella (bladderpod) is another native genus with excellent use-potential for seed oil (10). This genus ranges from northern Mexico to Wyoming and Tennessee to Arizona. While there are two distinct oil types, the species that contain predominantly C-20 hydroxy acids are of the most interest. Research emphasis has been on *L. fendleri* (Gray) Wats. and *L. gordonii* (A. Gray) S. Wats. Both species are widely distributed. *L. angustifolia* (Nutt.) is of interest because of larger seed and adaptation.

Of the more than 70 species, 20 have been field-tested. Progress with *L. fendleri* has been made in Arizona with respect to erectness and height but not to seed retention. Small plots have given acceptable yields, but these were not attainable on a larger scale. An increased level of developmental work will be necessary to advance this genus to possible crop status. The market potential for the oil is sufficiently high to justify extensive agronomic and breeding efforts.

3. *Epoxy Acid Seed Oils*

There are no natural sources of epoxy fatty acids used. The prime candidate, *Vernonia anthelmintica* (L.) Willd., was introduced from India and Pakistan (11). Yield potential and breedability were favorable. However, no genetic variability in seed retention has been identified. Our attention was diverted temporarily to *Euphorbia lagascae* Spreng. because of high seed oil content and good plant habit. Again, seed shattering, even worse than that of *V. anthelmintica*, was encountered.

In late 1970, *Stokesia laevis* (Hill) Greene, commonly re-

ferred to as Stokes aster, became a new crop candidate (12,13).
Small field plantings were initiated in 1971. Stokes aster is
an attractive perennial plant native to Georgia, Florida,
Alabama, Mississippi, and Louisiana. A wild population in
South Carolina has not been seen in recent years. Populations
in North Carolina appear to be naturalized garden-escapes.
This species is an excellent ornamental and is of limited
availability as seed and plants. There are distinctively
early and late blooming types, different flower colors and
forms, variations in leaf shapes, and various other morphological differences. In conjunction with our selection and cultural studies, superior ornamental types are being identified
and evaluated.

TABLE 2

Seed composition of 20 accessions of Stokesia laevis

Constituent	Range in composition (%)
Oil content	27-44
Epoxy acid	63-79
Free fatty acids	6-40

Data provided by Dr. R. Kleiman, Research Leader, USDA, ARS,
Northern Regional Research Center, Peoria, Illinois.

Seed oil composition of 20 accessions from the total collection of 30 is given in Table 2. These data support the
probability of selecting for seed oils as high or higher than
50%. Some yield estimates have been low, but selections with
superior seed retention show excellent promise for further
development. Variability in seed retentiveness abounds within
the collection.

Problems associated with stand establishment and diseases
are of more concern than yield potential or seed retention.
Seeds germinate very slowly in soil; hence, weeds become a
serious problem. Also, the time span from direct seeding to
heavy seed production is too long. Most of our plantings are
established with transplants. Germination tests, as well as
field observations, show us the severity of the establishment-
from-seed problem but give no leads to its solution. Different approaches will be tried to eliminate or partially alleviate this difficulty. Preliminary herbicide studies are underway.

This interesting cross-pollinated plant attracts a wide

range of insects to its showy flowers, which may reach a diameter up to 4 1/2 inches. By utilizing early and late blooming types, flowering can be sustained from mid-June until
frost.

The deficiencies associated with Stokes aster are not
considered insurmountable. The plant shows promise of becoming a natural source of epoxy fatty acid, and at the same time,
adding beauty to the landscape as well as food for many
insects.

4.   *Source of rotenone*

Tephrosia vogelii Hook. provides a potential leaf source
of rotenone (14). Because of program limitations, sufficient
resources could not be allocated to solve the problems of low
rotenone content, poor seed production, and breeding problems
in *T. vogelii*. Seed production requires a tropical (frost-
free) location. Good vegetative yields were produced in
Maryland, North and South Carolina, and Puerto Rico.

II.   DISCUSSION AND FUTURE PROSPECTS

Several new crops hold promise for commercial acceptance.
Kenaf is on the verge of possible crop status. *Limnanthes*,
Lesquerella, and *Stokesia* are native plants with readily
accessible germplasm. We are attempting to introduce their
dulture. Sustained cooperative efforts are needed to successfully introduce a new crop. More emphasis on breeding, including availability of germplasm in early stages of development,
would enhance the program. Flexibility is a key ingredient in
recognizing problems, identifying special use properties, and
unusual new leads, and seeing potential dead ends or extremely
long-range solution of problems. Obviously, the number of
species involved needs to be manageable within program resources. Effective coordination compatible with administra-
tive structure, provides program direction and unity.

The United States has had to depend on plant introductions
for most of its crop species. The plant kingdom is large,
variable, and pliable; it should continue to yield new crops
for food, feed, fiber, industrial uses, medicinal purposes,
and environmental enrichment to the benefit of man. Let's
continue the search, and at the same time, preserve the germplasm and information gained along the way.

III.   REFERENCES

1.   White, G. A., Willingham, B. C., Skrdla, W. H., Massey, J.
     H., Higgins, J. J., Calhoun, W., Davis, A. M., Dolan, D.
     D., and Earle, F. R., *Econ. Bot.* 24, 22 (1970).

2. Adamson, W. C., Stone, E. G., and Minton, N. A., *Crop Sci.* 14, 334 (1972).

3. White, G. A., Clark, T. F., Craigmiles, J. P., Mitchell, R. L., Robinson, R. G., Whiteley, E. L., and Lessman, K. J., *Econ. Bot.* 28, 136 (1974).

4. White, G. A., Cummins, D. G., Whiteley, E. L., Fike, W. T., Greig, J. K., Martin, J. A., Killinger, G. B., Higgins, J. J., and Clark, T. F., "Cultural and harvesting methods for kenaf", U.S. Dept. Agr. Prod. Res. Rept. 113, 1970.

5. White, G. A., and Haun, J. R., *Econ. Bot.* 19, 175 (1965).

6. Lessman, K. J., and Meier, V. D., *Crop Sci.* 12(2), 224 (1972).

7. White, G. A., and Higgins, J. J., "Culture of crambe--a new industrial oilseed crop" U.S. Dept. Agr. Prod. Res. Rept. 95, 1966.

8. Higgins, J. J., Calhoun, W., Willingham, B. C., Dinkel, D. H., Raisler, W. L., and White, G. A., *Econ. Bot.* 25, 44 (1970).

9. Gentry, H. S., and Miller, R. W., *Econ. Bot.* 19, 25 (1965).

10. Barclay, A. S., Gentry, H. S., and Jones, Q., *Econ. Bot.* 16, 95 (1970).

11. Berry, C. D., Lessman, K. J., White, G. A., and Earle, F. R., *Crop Sci.* 10, 178 (1970).

12. Gunn, C. R., and White, G. A., *Econ. Bot.* 23, 130 (1974).

13. White, G. A., and Gunn, C. R., *Garden Jour.* 24(3), 84 (1974).

14. Gaskins, M. H., White, G. A., Martin, F. W., Delfel, N. E., Ruppel, E. G., and Barnes, D. K., "*Tephrosia vogelii* Hook. f. A source of rotenoids for insecticidal and piscicidal use", U.S. Dept. Agr. Tech. Bull. 1445, 1972.

NONFOOD USES FOR COMMERICAL VEGETABLE OIL CROPS

E. H. Pryde

Northern Regional Research Center
Agricultural Research Service
U.S. Department of Agriculture
Peoria, Illinois 61604

I. INTRODUCTION

In their introduction to *Materials: Renewable and Nonre-newable Resources*, Abelson and Hammond (1) state that renewable resources are crucial to an enduring human civilization and thet the United States apparently has not yet got its priorities straight on this point. All supplies of chromium, manganese, cobalt, tin, columbium, aluminum, titanium, plantinum and palladium, and more than half of out fluorine, mercury, bismuth, nickel, selenium, zinc, tungsten and cadmium are imports (2). About one-half of our oil requirements is imported, and this fraction will be increasing. Only the enormous agricultural strength of the United States prevents a persistent deficit in our balance of payments (3).

Certainly, renewable resources are present in our economy, but the major question is to what extent can they replace nonrenewable ones. Our renewable forest resources offer a major alternative to petroleum and even to coal for chemical intermediates and plastics (4), and so do also, to a lesser degree, corn (5,6) and vegetable oils (7).

However, Landsberg (2) points out that judgments will be necessary on the desirability of such substitutions. Land devoted to growing industrial materials, for example, would not be available for raising food crops. In order to increase productivity, increased fertilization would be necessary. Environmental problems will emerge. Further, many basic material industries increasingly tend to resist technological innovation, although there is no lack of new inventions or ideas (3). Companies are faced with unmanageable capital replacement problems plus the costs of compliance with health, safety and environmental regulations.

Obviously a national policy for materials is needed. Huddle (3) has reviewed past developments in materials policy and makes the following recommendations. First, we should identify those economic, technological and political elements that encourage or discourage good materials management and respond to these findings. Second, we should maintain surveillance of all aspects of national materials management to detect emerging obstacles to good performance and lags in execution. Third, we should develop an institutional capability for (i) quickly correcting sudden deficiencies in national materials management and (ii) detecting and gradually correcting trends in the management of the materials cycle.

In the last analysis, however, the problem becomes one of economics, if we can no longer afford to pay the price for imported materials or the high cost of extracting low-grade ones, then less expensive alternatives will have to be found or we will be doing without.

The automotive industry is an interesting example of rapidly changing materials usage. By 1980, plastics will be on the verge of replacing cast iron as the second largest component of the average automobile (8). Plastics originally entered this market because of their lower cost, but they have become most important in the drive to effect improved fuel economy by weight reduction. Under these conditions, cost per unit property of material becomes extremely important (9).

The increasing use of plastics raises questions about environmental impact and energy requirements. In one study on packaging materials, the Midwest Research Institute found, in the majority of cases investigated, that plastics had the advantage over nonplastics in both respects (10).

It is the intent of this paper to review briefly the vegetable oil industry, its relation to the total fats and oils industry and finally the present and possible future uses of industrial products derived from vegetable oils.

II. THE FATS AND OILS INDUSTRY

The magnitude of the fats and oils industry is easy to overlook, yet its more than 20 billion pounds amounts to one-fifth of the total petrochemicals industry. Vegetable oils, mainly soybean oil, contribute the greatest share to the industry (Table 1) (11). Soybean oil has a remarkable and glamorous history, having started from small beginnings of 1 million-pound production in the 1920's to the 9 billion pounds produced in the 1973/74 crop year and the record 9.5 billion pounds produced in 1975/76. In the United States, soybean oil supplied 76% of vegetable oil food requirements and 67% of the total edible fat requirements in 1974.

TABLE 1

U.S. Production and Consumption of Fats and Oils, 1975 (11)
(Million Pounds)

Fat or Oil	Production	Exports	Consumption Edible Uses	Consumption Inedible Uses
Vegetable	9,811	1,300	9,200	649
Industrial	868	136	---	572
Animal	6,169	1,116	833	2,968
Marine animal	---	3	---	52
Other	---	---	20	33
Total	16,848	2,555	10,053	4,274

About two-thirds of total fats and oils production goes into food uses and one-third into nonfood uses--an annual market of 4-5 billion pounds. This ratio of food to nonfood uses has been maintained at about 2:1 for the past 40 years (Table 2) (12). Per capita consumption for both food and nonfood fats has increased modestly since 1935-1939 and consumption of food fats may reach 60 pounds per capita by 1980 (13-15).

TABLE 2

Per Capita Consumption of Fats and Oils, 1935-1975 (12)

Period	Average Consumption (pounds per capita) All Industrial Products	Average Consumption (pounds per capita) All Food Products	Ratio, Food: Nonfood Uses
1935-39	22.2	45.4	2.05
1950-59	24.0	44.9	1.87
1960-69	25.8	48.0	1.86
1970-75	24.8	53.4	2.15

There appears to be more possibility for growth in the nonfood category as the concept grows for exploiting our renewable in place of nonrenewable resources.

Whereas vegetable oils predominate in food uses, animal fats predominate in nonfood uses (Table 1).

Nevertheless, food oils contribute 15% and industrial oils contribute 17% to the total consumption of fats and oils

in inedible uses. If the 1149 million pounds of inedible tallow consumed in animal feed preparation are subtracted, these
respective percentages increase to 20% and 23% of the remaining 3494 million pounds employed in industrial uses (Table 3).

TABLE 3

Relative Contributions of Fats and Oils
to Edible and Inedible Uses

Fat or Oil	Contribution, %		
	Edible	Inedible	Inedible[a]
Food vegetable oils	89	15	20
Animal fats	11	66	55
Industrial vegetable oils	--	17	23
Marine animal oils	<0.3	1	1
Other	--	1	1
	100	100	100
Total, million pounds	10,137	4,643	3,494

[a]Exclusive of inedible tallow going into animal feeds.

III. VEGETABLE OILS AND DERIVED FATTY ACIDS

A. Utilization for Industrial Products

The vegetable oils and related products that are important to industrial utilization are listed in Table 4 (11).
Oils that are used exclusively for industrial products include
castor, linseed, tall oil, fatty acids, tung, and vegetable
oil foots. Edible oils that are also used to varying degrees
for industrial products include coconut, cottonseed, palm,
peanut, safflower, and soybean oils. Surprisingly, soybean
oil contributed the greatest volume to industrial utilization
for 1973, followed by coconut oil, tall oil fatty acids, linseed oil, and castor oil, in that order. Exports for these
oils are listed also in Table 4, because, undoubtedly, some of
the exported oils go into industrial products. Such products
have been summarized for the 3 billion pounds of fats and oil
products consumed as chemical intermediates in the European
Common Market (17). About 1200 million pounds of oils were
imported in 1973.

The dominant position occupied by soybean oil in both
food and nonfood oils has already been described. Of all other
edible oils, coconut oil occupies the next most important position with its 57% going into industrial products. Safflower

oil occupies third position with its 25% going into industrial use.

TABLE 4

Domestic Utilization of Vegetable Oils for Industrial Products and Exports for 1975

Vegetable Oil	Domestic Disappearance for Industrial Products, 1975 (11)		Exports (million lbs.)
	Amount (million lbs.)	Percent of Total Domestic Disappearance	
Castor	87	100	68[a]
Coconut (14)	426	49	869[a]
Corn	---	---	45
Cottonseed	10	1	657
Linseed	130	100	87
Palm	10	1	960[a]
Palm kernel	---	---	163[a]
Peanut	3	3	27
Safflower	17	31	---
Soybean	184[b]	3	758
Sunflower	---	---	---
Tall oil fatty acids (15)	242	100	23
Tung	31	100	30[a]
Vegetable oil foots	92	100	26

[a] Imported
[b] Dropped from an estimated 325 million pounds in 1974, and 508 million pounds in 1973, probably because of high prices that followed the oil embargo (16). This figure includes soybean foots and loss, but does not include epoxidized soybean oil.

Of the industrial oils, only linseed oil is now produced to any substantial amount in this country. In times of emergency, linseed oil has served as an edible oil in some parts of the world. Linseed oil supplies have suffered a severe decline--from more than 1400 million pounds in 1950 to about 208 million pounds in 1975. The major reason for the decline can be ascribed to latex paints made from cheap petrochemicals.

B. Oil Composition

Because they have a direct bearing on the end use, compositions of some oils are listed in Table 5. Noteworthy are the characteristic acids in some of the oils, including (per-

TABLE 5

Composition of Vegetable Oils

Oil	Saturated Fatty Acid, %		Unsaturated Fatty Acid, %			Other or Characteristic Fatty Acid
	Lower	C-18	Oleic	Linoleic	Linolenic	
Castor	1	1	3	4	<1	Ricinoleic (89%)
Coconut	88	4	5	3	--	Lauric (48%)
Corn	12	2	27	59	1	---
Cottonseed	26	3	17	53	--	---
Crambe	2	<1	17	9	6	Erucic (57%)
Linseed	6	4	19	15	57	---
Oiticica	7	5	6	--	--	α-Licanic (78%)
Olive	17	3	62	15	1	---
Palm	48	4	38	10	--	Palmitic (46.8%)
Peanut	11	2	51	31	--	---
Rapeseed	4	1	17	13	5	Erucic (46%)
Rapeseed	5	2	60	17	7	Eicosenoic (3%)
Safflower	6	4	14	76	--	---
Safflower	5	2	81	12	--	---
Soybean	11	3	22	55	8	---
Sunflower	7	3	14	75	--	---
Tall oil fatty acids	--	3	48	37	--	Conjugated (6%)
Tung	4	1	8	4	3	α-Eleostearic (80%)

cent of total fatty acids) ricinoleic (89%) in castor, lauric
(48%) in coconut, erucic (57%) in crambe, linolenic (57%) in
linseed, α-licanic (78%) in oiticica, palmitic (47%) in palm
and α-eleostearic (80%) in tung oils.

Ricinoleic (12-hydroxy-9-octadecenoic) acid is unique in
being the starting material for two valuable industrial intermediates: sebacic acid by alkali fusion and 1-undecenoic acid
by pyrolysis. Lauric (dodecanoic) acid is an industrial intermediate in the preparation of a wide variety of surface-active
agents. Erucic (13-docosenoic) acid is valued for its long,
C-22 aliphatic chain, which reduces volatility and increases
hydrophobic properties in its compounds. It potentially is a
source of the C-13 dibasic acid, brassylic acid. Oils with
three double bonds in the fatty acid chain as in linolenic
(cis-9, cis-12, cis-15-octadecatrienoic) acid, α-licanic (4-
keto-cis-9, trans-11, trans-13-octadecatrienoic) acid and
α-eleostearic (cis-9, trans-11, trans-13-octadecatrienoic)
acid are used as pigment binders in the paint industry because
of their rapid drying properties. Linseed oil with its 57%
linolenic acid content is by far the most important.

IV. VEGETABLE OILS AS RENEWABLE RESOURCES

A. The Problems

The grand total of synthetic organic chemicals production
in 1974 amounted to 184 billion pounds for all categories, of
which plastics and resins is one of the largest at 30 billion
pounds (18). The great preponderance of these chemicals is
from petroleum. Faced with declining U.S. petroleum production and increased prices for imported oil, to what extent can
the chemical industry turn towards the fats and oils industry?
the latter already supplied 4.6 billion pounds (3%) to its
needs. The question raises several others in turn, for example:
1. Will all agricultural production be required for food
    purposes, or will land be available for nonfood crops?
2. What are the energy requirements for producing agricul-
    tural crops?
3. What are the economics of using agricultural products as
    chemical intermediates?
It is not the intent of this paper to answer these questions
in depth, but some indications of probable trends are possible.
It should be noted that vegetable oils are probably the
only economically feasible alternatives to petrochemicals as
sources for long-chain, aliphatic compounds.

B. Land Requirements

At the last census, about 330-million acres of cropland

TABLE 6

Land Utilization (United States) (19)
(Million Acres)

	1950	1960
Land in farm		
Cropland used for crops	387	333
Cropland idle or in cover crops	22	51
Cropland used only for pasture	69	88
Grassland pasture	417	452
Forest and woodland	221	112
Farmsteads, roads	46	28
Land not in farms		
Grazing land	402	287
Forest land not used for grazing	368	475
Other (cities, railroads, wasteland)	341	438
Grand Total	2,273	2,264

in the United States was used for crops (Table 6) (19), representing about 10% of the world's 3.3-billion acres now tilled. It has been estimated that as much as 7.8-13 billion acres could be made available in the world (20,21). Canada and the United States together contribute about 90% of world exports in food products, but these shipments provide only about 7% of the food requirements in the underdeveloped world (22).

In the 50 states, only about a fifth of all land is cropland. Livestock grazing requires well over a third of all land, and forest land occupies another third (23). Distribution of acreage between food and feed grains in 1974 was as follows (24):

Feed grains	100,717,000 acres
Food grains	68,925,000 acres

Harvested acreage for selected crops in the United States for 1974 indicates corn, wheat, hay, soybeans, and sorghum in that order, to be our most important crops (Table 7) (25).

If soybean oil were to become the sole intermediate to 180 billion pounds of synthetic organic chemicals, then some 600 million acres of soybean plantings would be required in contrast to the present 54 million acres—a possible but unlikely event since the American diet then would consist mainly of soybean protein. Furthermore, soybean oil could not possibly meet many needs, particularly where aromatic petrochemicals are concerned. For 30 billion pounds of plastics, 100 million acres of soybeans would be required in addition to the present acreage.

   Admittedly, such large soybean acreage increases would
use farm cultural energy more efficiently than meat production.
It has been estimated that 38 gallons of oil are required to
produce the 667 pounds of animal products per annum per capita
in contrast to the 14 gallons of oil required to produce the
787 pounds of plant products consumed in 1972 (26). Furthermore, the animal itself is an inefficient machine for producing food, the actual conversion of feed to food energy varying
by a factor of 0.05 for beef, 0.13 for eggs and pork, to 0.19
for dairy products (27).

TABLE 7

*Harvested Acreage of Selected Crops
in the United States, 1974*

Crop	Acreage Harvested 1,000 Acres (25)
Corn	76,766
Wheat, all	65,459
Hay	60,546
Soybeans	52,460
Sorghum	16,834
Oats	13,325
Cotton	12,547
Barley	8,281
Rice	2,569
Flaxseed	1,645
Peanuts	1,472

   A more realistic prediction is a 20% increase in soybean
production to 1,835 million bushels by 1985 despite increased
competition from Brazilian soybeans and rising palm oil imports (28). This increased production is estimated to require
a 13% increase over the 1975 harvested acreage of 53.6 million
acres, assuming a yield increase to 30.25 from 28.4 bushels
per acre. The rise in soybean production will result from
increased demand for soybean meal as a livestock and poultry
feed and from increased exports.

   It is likely that soybean oil will be in surplus, but how
much more will be used for industrial products is unpredict-
able and will depend upon price. Too high a price will discourage industrial use; too low a price will likely encourage

encourage the farmer to plant other crops.

Furthermore, it is quite reasonable to expect that land will be available for growing needed industrial oils. There is substantial evidence that the world could attain a favorable food production-demand balance within the next 3 decades (20).

C. Production Costs

In 1970, an estimated 12-13% of all U.S. energy was consumed for food production, processing, preservation, and cooking (29). The total energy consumed in these areas was divided up as follows (Table 8) (30):

Farm costs	24%
Processing	39
Commercial and residential refrigeration and cooking	37

It is interesting to note that farm costs require the least expenditure of energy in the three categories.

TABLE 8

Energy and Agriculture (30)
(10^{12} kcal)

Component	Year			
	1940	1950	1960	1970
Farm	125	303	374	526
Processing industry (including packaging and transport)	286	454	572	842
Commercial and home (refrigeration and cooking)	275	377	495	804

Costs of producing selected crops in the United States have been estimated for the 1974 growing season (31,32). A comparison of cost items of high-energy input shows that fertilizer is most important for corn, while fuel and herbicide are most important for soybeans, fertilizer and insecticide for peanuts (Table 9) (31). In terms of dollars per acre, the total direct investment required decreases in the order peanuts, corn, soybeans, winter wheat, and flax (Table 10). In

TABLE 9

Cost Items of High Energy Input, 1974 (31)

Crop	Total Direct Cost, Dollars Per Acre	Cost Item, Dollars Per Acre			
		Fuel	Fertilizer	Insecticide	Herbicide
Corn	103.55	9.66	38.81	7.62	1.58
Flax	33.33-42.01	4.40-5.59	0.73-3.59	---	0.55-0.65
Peanuts	215.08	15.21	28.07	34.71	14.54
Soybeans	53.18	6.65	4.79	0.43	6.40
Winter wheat	42.89	4.40	8.05	0.43	0.26

TABLE 10

Crop Production Costs, 1974 (31)

| Crop | Total Direct Costs, Exclusive of Land | |
	Dollars Per Acre	Dollars Per Unit
Cron	97.43	1.31/bu
Flax	33.33-42.01	5.09-3.79/bu
Peanuts	215.08	7.19/100 wt
Soybeans	53.18	2.16/bu
Winter wheat	42.89	1.45/bu

terms of dollars per bushel, the total of all costs decreases
in the order peanuts, flax, corn, soybeans, and winter wheat
(Table 11). In 1974, a relatively poor growing season, and
for selected areas of the United States, it appears that the
soybean was a profitable crop for the United States farmer.

TABLE 11

Average Yields, Costs and Prices
Received by Farmers, 1974 (31)

Crop	Average Yield bushels/acre	Total of All Costs,[a] dollars/bushel	Average Price Received by Farmers, dollars/bushel
Cron	82.2	2.39	2.95
Flax	6.6-11.0	4.00-6.37	9.45
Peanuts	2,990[b]	8.87[c]	0.179[d]
Soybeans	24.7	2.79	6.69
Winter wheat	29.5	1.89	3.88

[a]Exclusive of land costs.   [c]Dollars per hundred weight.
[b]Pounds.   [d]Dollars per pound.

Regarding balancing energy and food production in the
far future (2050-2200 A.D.), Chancellor and Goss (33) are optimistic but state that requirements for such a balance include zero population growth and provision of all energy by
the sun.

However, in examining energy needs and food yields,
Heichel (34) concludes that modern agriculture uses cultural

energy 100- to 500-fold more efficiently than plants use sun-
light and 10- to 50-fold more efficiently than animals metabo-
lize feed. Gains in energy efficiency are most likely to be
realized by developing more efficient or new cropping systems.
Further gains could be made by genetic improvements in photo-
synthetic efficiency, use of crops requiring little processing,
more extensive use of plants as protein and energy sources,
and by frugal cultural practices.

D.   Prices

The availability of land, the relatively low production
costs for producing vegetable oils and other factors suggest
that vegetable oil prices may remain at fairly low levels compared to escalating prices for petrochemicals. In fact, a
comparison of petrochemical and vegetable oil prices reveals
that the former are fast catching up and may eventually exceed
the latter (Table 12). Under such circumstances, vegetable
oils become attractive as raw materials for industrial products.

TABLE 12

Prices for Fats and Oils vs. Petrochemicals

| Material | Price, Dollars Per Pound | | |
	1972	June 14, 1976	Future
Petrochemical			
Ethylene	0.030	0.12	0.20 (1980)
Propylene	0.029	0.09	0.11-0.16 (1976)
Butadiene	0.10	0.22	0.30 (1984)
Benzene	0.028	0.11	0.14 (1977)
Fat or oil			
Coconut	0.098	0.19	0.16 (est)
Linseed	0.089	0.24	---
Palm	0.106	0.17	0.17 (est)
Soybean	0.113	0.17	0.17 (est)
Tallow, inedible	0.060	0.15	<0.15

V.   INDUSTRIAL MARKETS

A.   Coatings

The coatings industry requires large amounts of unsaturated vegetable oils and this is the main outlet for these oils
(Table 13) (35). Consumption has dropped by about 40% from
the 1950's to the 1970's, and the reason can be ascribed to

TABLE 13

Consumption of Vegetable Oils in the Drying Oil Industries

Oil	Annual Consumption in Drying Oil Industries, Million Pounds (35)		
	1950–1959	1960–1969	1970–1973
Castor	58	87	45
Linseed	515	331	259
Oiticica	11	9	---
Soybean	202	187	150
Tall oil	104	103	75
Tung oil	56	34	28
All oils	1,059	862	628

cheap petrochemicals used in latex paints. However, several properties of unsaturated vegetable oil products in coatings are such that they cannot be easily obtained by petrochemical products, and the result is that vegetable oils retain an estimated 40% of the paint binder marker.

B.   Plasticizers

Plasticizers, mainly for vinyl plastics, are another large outlet for fatty derivatives (Table 14) (36). Epoxidized soybean oil, a combination plasticizer-stabilizer for poly(vinyl chloride), is one of the most important industrial applications for this oil. An estimated 15% of the total plasticizer market is estimated to be based on fatty-derived materials.

The major portion of the market is occupied by phthalate esters. These have become ubiquitous environmental contaminants, and there is basis for questioning the safety of such materials (37).

C.   Surface Active Agents

By far the largest use for fatty-derived materials is in surface-active agents (Table 15) (38). Almost half of the total market goes to fatty materials, mostly tallow but also considerable amounts of vegetable oil, fatty acids. The short chain acids, particularly the 12-carbon lauric acid, in coconut oil are especially useful in many detergent applications.

The properties of the straight-chain aliphatic compounds from vegetable oils and animal fats are so desirable that the petrochemical industry has synthesized mixtures of fatty

TABLE 14

Vegetable Oil-Derived Plasticizers (36)

Plasticizer		1974 Production, Thousand Pounds
Di(2-ethylhexyl) azelate		9,262[a]
Epoxidized esters, total		153,923
Epoxidized soya oils	126,946	
Octyl epoxytallates	14,937	
Glyceryl monoricinoleate		108
Isopropyl myristate		4,997
Isopropyl palmitate		7,444
Oleic acid esters, total		12,628
Sebacic acid esters, total		7,665
Stearic acid esters, total		15,164
Others[b]		122,583
Total		333,774
All plasticizers		1,891,685

[a]Sales for 1973.
[b]Includes data for azelaic, citric and acetylcitric, lauric, myristic, palmitic, pelargonic and ricinoleic acid esters, glycerol and glycol esters and other acyclic plasticizers, not separately shown.

alcohols, either by telomerization of ethylene by a Ziegler catalyst or by carrying out the oxo reaction on 1-alkenes.

D. Polyamides

Vegetable oil products go into several different types of polyamides, both high-molecular weight polymers and low-molecular weight resins. Nylons 11 and 610 are examples of the former, and dimer acid polyamides and reactive polyamides are examples of the latter.

Ricinoleic acid produces either undecylenic or sebacic acid depending upon reaction conditions. Undecylenic acid is the starting material for nylon 11, a low-moisture-absorbing, high-molecular-weight polyamide that is used for engineering thermoplastics, powder coatings for metal furniture and textile applications. Sebacic acid is a component of nylon 610 used for monofilaments, bristles, and various plastics applications. Sebacic acid is receiving stiff competition from petrochemically derived dodecanedioic acid as a component of polyamides. Data are not available on production statistics

TABLE 15

Fatty Acid-Derived Surface Active Agents (38)

Fatty Compound	1974 Production, Million Pounds	1974 Unit Value, Dollars per Pound
Acids and salts	838.7	0.21-1.37
Acid esters	262.2	0.52
Alcohol sulfates	212.5[a]	0.37
Alcohols, alkoxylated	605.6[b]	0.27
Amines and quaternary ammonium salts	283.8	0.63
Amine condensates with diethanolamine	89.1	0.65
Total	2,291.9	
All surface active agents	4,696.5	0.30

[a]Sales for 1973.
[b]Includes about 400 million pounds of alkoxylated mixed linear alcohols.

for these materials, but a fair estimate might be that the fatty-derived materials take up about 15% of the total 150 million pound and growing nylon plastics market, in which the dominant polyamide is nylon 6.

Dimer acids are made by polymerization of tall oil or soybean fatty acids. Dimer acid polyamides make excellent hot melt adhesives and flexographic printing inks for a total market of about 25 million pounds. Reactive polyamides from dimer acid and polyethylene amines cure epoxy resins for adhesives and coatings to make up another 25-million-pound market.

E. Lubricants

Lubricants are of many different types, but those important to fatty materials include extreme pressure and transmission fluids from sulfurized oils (50-60 million pounds/year), jet engine ester lubricants (30 million pounds/year), grease additives as lithium 12-hydroxystearate (30 million pounds/year), and crankcase and other oils from synthetic ester lubricants (30 million gallons/year).

Sulfurized sperm oil was uniquely suitable for transmission fluids, but since the sperm whale has become an endangered species, other fatty esters have been required. Jet engine

lubricants make use of pelargonic acid, obtained from the oxi-
dative cleavage of oleic acid.   12-Hydroxystearic acid is obtained by hydrogenation of ricinoleic acid from castor oil.
Castor oil itself can be used as a lubricating oil in some
applications.
    Synthetic ester lubricants may be based on a dicarboxylic
acid such as adipic, azelaic or sebacic acids.   The last two
are fatty derived, but adipic is from petrochemicals.   Synthetic ester lubricants can also be based on polyol esters such
as mixed short-chain fatty acid esters of pentaerythritol.
Synthetic ester lubricants have great potential becuase of improved fuel consumption, reduced wear, longer life, and improved low-temperature performance compared to ordinary lubricants.

F.   Lubricant Additives

    Lubricant additives are essential as processing aids for
several types of plastics and include such materials as
ethylene *bis*-stearamide, oleamide, erucamide, various fatty
esters, and metallic stearates.   The total market for such
additives is estimated to be on the order of 50 million pounds
per year.   Erucamide is particularly valuable as a slip or
antiblock agent for polyethylene film.

G.   Glycerol

    Glycerol is made from both natural and synthetic sources,
with the latter supplying 68% of total glycerol capacity.
Either hydrolysis or saponification of vegetable oils and animal fats yields glycerol as a valuable byproduct.   It is also
formed by methanolysis of coconut oil, which is the first step
in making short-chain fatty alcohols.
    Synthetic glycerol capacity is rated at 345 million
pounds and natural glycerol capacity at 165 million pounds per
year.   Uses of glycerol are distributed as follows: drugs and
cosmetics (19%), exports (18%), alkyds (17%), tobacco (12%),
food and beverage (10%), cellophane (10%), urethane polyols
(7%), explosives (4%) and miscellaneous (3%) (39).

H.   Lecithin

    An important byproduct from the refining of crude vege-
table oils is lecithin, contained in crude soybean oil to the
extent of 2-3% (40).   A variety of lecithin grades are available, and these are used extensively as emulsifiers for many
different food products including bakery goods and chocolate
(41).   World demand for lecithins has been estimated to be on
the order of 100,000 tons per year (40).

I.   Proteins

    Shortly before World War II, Henry Ford succeeded in us-
ing the protein in soybean meal in conjunction with phenol-
formaldehyde resins to make such automotive parts as distribu-
tor caps, horn buttons and window frames. During the war,
this practice was stopped, and it has never been revived,
mainly because of the poor hydrolytic stability of the molded
parts and because of the need for soy protein as an animal
feed.
    Soybean meal at one time was used quite extensively as a
polywood adhesive, but this market has declined drastically
over the past 20 years to a present consumption of 4-5 million
pounds per year. Soy protein isolate, containing 90-95% protein compared to the 50-54% in meal has maintained better markets and is consumed to the extent of 30-60 million pounds
annually as an adhesive for paper coating pigments.

VI.   FUTURE MARKETS

A.   The Need for Research

    That fatty-derived industrial products are essential to
the economy of the United States is obvious. It is probable
that industrial demand will increase in future years and that
ample supplies will be available for such products. Availability will be due to several reasons. It is doubtful that
demand for edible oils will keep pace with that for soy protein. Furthermore, edible oils are receiving increasing down-
ward pressure on price because of rapidly expanding soybean
and palm oil production in foreign countries.
    Although the size of the fats and oils industry is
roughly 1/5 of that for petrochemicals, research investment in
the two industries is probably far from maintaining that ratio.
The probable reason for a disparity is that, while petrochemicals represent only 5% of the total volume of the petroleum
industry, the petrochemical industry had a wide financial base
to operate from. Heavy investment in petrochemicals research
has paid off handsomely. In contrast, smaller vegetable oil
processing companies have not had the capital for comparable
research investment. Further, investments of time and effort
in health, safety, and environmental concerns have left little
for applied and developmental research.
    Research of all types, including basic, applied, develop-
mental, pilot demonstrations and the like, is needed to lay
the foundations for future technologies, to diversify and expand markets for fats and oils and to transfer petrochemical
technology, where applicable, to the fats and oils industry.
    It becomes imperative to obtain optimum utilization for
vegetable oils as valuable, renewable resources both for do-

mestic consumption and for value-added products for export.

B.   USDA Utilization Research

The Regional Research Centers of the Agricultural Research Service, U.S. Department of Agriculture, were established by Congress in 1938 to seek new and expanded uses for farm products. The Northern Regional Research Center was given responsibility for research on soybean and linseed oils as well as corn, wheat, and grain sorghum. The Eastern Regional Research Center was given responsibility for research on animal fats as well as dairy products, meat, hides, fruits and vegetables. These two laboratories have carried out the bulk of research on fatty derivatives, but the Southern Center has done research on cottonseed, peanut, and tung oils; the Western Center, on safflower and castor oils; and the Southeastern Center, on sunflower oil.

Developments of these laboratories that have attained industrial acceptance and that have contributed to products now widely used include the following:
  -epoxidized soybean oil as a plasticizer/stabilizer for
   vinyl plastics
  -dimer acid polymers for adhesives, coatings and printing
   inks
  -aqueous linseed emulsions for curing and protecting con-
   crete highways
  -rigid urethane foams from castor oil
  -tallow-based soaps contaiining a lime-soap dispersing
   agent.

Linseed emulsion paints were available to the consumer for a short period, but these soon lost out to lower cost, petrochemically derived latex paints.

Developments from the Regional Research Center that have not yet been placed into commercial operation but that perhaps should now be reconsidered by industry include the following:
  -coatings from fatty-vinyl ether polymers
  -lubricants from cyclic and carboxylated fatty acids
  -plasticizers with improved properties from
   polyacetoxymethylated oils and brassylate esters
  -nylon 9 and nylon 1313 plastics
  -miscellaneous industrial products from isopropenyl esters
  -food emulsifiers from glycerol glucoside esters.
Some of these have been described in greater detail in a recent publication (42).

VII. REFERENCES

1.   Abelson, P. H., and Hammond, A. L., in "Materials: Renew-
     able and Nonrenewable Resources," p. vi, Amer. Assoc. for
     the Adv. of Sci., Washington, D.C., 1976.

2. Landsberg, H. H., in "Materials: Renewable and Nonrenew-
    able Resources," p. 1, Amer. Assoc. for the Adv. of Sci.,
    Washington, D.C., 1976.
3. Huddle, F. P., in "Materials: Renewable and Nonrenewable
    Resources," p. 18, Amer. Assoc. for the Adv. of Sci.,
    Washington, D.C., 1976.
4. Goldstein, I. S., in "Materials: Renewable and Nonrenew-
    able Resources," p. 179, Amer. Assoc. for the Adv. of
    Sci., Washington, D.C., 1976.
5. Moore, C. A., Reprint of Papers, Div. of Chem. Mark. and
    Econ., p. 73, Amer. Chem. Soc. Meeting, Philadelphia,
    Pennsylvania, April 7-9, 1975.
6. Otey, F. H., Reprint of Papers, Div. of Chem. Mark. and
    Econ., p. 87, Amer. Chem. Soc. Meeting, Philadelphia,
    Pennsylvania, April 7-9, 1975.
7. Pryde, E. H., Gast, L. E., Frankel, E. N., and Carlson,
    K. D., Reprint of Papers, Div. of Chem. Mark. and Econ.,
    p. 100, Amer. Chem. Soc. Meeting, Philadelphia, Pennsyl-
    vania, April 7-9, 1975.
8. Anon. *Chemical Marketing Reporter*, p. 17, April 5, 1976.
9. Deanin, R. D., and Driscoll, S. B. *Mod. Plast.*, 110,
    April 1973.
10. Anon. *Mod. Plast.*, 56, April 1974.
11. U.S. Bureau of the Census. Current Industrial Repts.,
    Fats and Oils, Production, Consumption, and Factory and
    Warehouse Stocks, Series M20K(74)-13, Washington, D.C.,
    1975.
12. Fats and Oils Situation, FOS-283, p. 16, Econ. Res. Ser.,
    U.S. Dept. Agr., Washington, D.C., 1976.
13. Boutwell, W., Doty, H., Hacklander, D., and Walter, A.,
    Analysis of U.S. Fats and Oils Industry to 1980. U.S.
    Dept. Agr. Econ. Res. Ser. Publ. No. ERS-627, May 1976.
14. Fats and Oils Situation, FOS-281, p. 12, Econ. Res. Ser.,
    U.S. Dept. Agr., Washington, D.C., 1976.
15. Fatty Acid Production, Disposition and Stocks Census and
    Pulp Chemical Association Statistics for Tall Oil Fatty
    Acids, Fatty Acid Producers' Council, New York, 1975.
16. Fats and Oils Situation, FOS-280, p. 10, Econ. Res. Ser.,
    U.S. Dept. Agr., Washington, D.C., 1975.
17. Ploog, Von U., and Reese, G. *Chem-Ztg.* 97, 342 (1973).
18. Synthetic Organic Chemicals. United States Production
    and Sales, 1974. U.S. Intl. Trade Com., ITC Publ. 776,
    p. 3, U.S. Govt. Printing Office, Washington, D.C., 1976.
19. Agricultural Statistics 1975, p. 420, U.S. Dept. Agr.,
    U.S. Govt. Printing Office, Washington, D.C., 1975.
20. Blakeslee, L. L., Heady, E. O., and Framingham, C. F.,
    "World Free Production, Demand, and Trade," Iowa State
    Univ. Press, Ames, Iowa, 1973.
21. Kottman, R. M. As quoted in *Farm J.* 99, A1, 1975.

22. Kastens, M. L. *Chem. Technol.*, p. 675, November 1975.
23. Our Land and Water Resources. Misc. Publ. No. 1290, Econ. Res. Ser., U.S. Dept. Agr., U.S. Govt. Printing Office, Washington, D.C., 1974.
24. Agricultural Statistics 1975, p. 434, U.S. Dept. Agr., U.S. Govt. Printing Office, Washington, D.C., 1975.
25. Agricultural Statistics 1975, p. 435, U.S. Dept. Agr., U.S. Govt. Printing Office, Washington, D.C., 1975.
26. Heichel, G. H. *Am. Sci.* 64(1), 64 (1976).
27. Heichel, G. H., Frink, C. R. *J. Soil Water Conserv.* 30, 48 (1975).
28. Hacklander, D. Fats and Oils Situation, FOS-281, p. 30, February 1976, Econ. Res. Serv., U.S. Dept. Agr., Washington, D.C.
29. The U.S. Food and Fiber Sector: Energy Use and Outlook. Econ. Res. Serv., U.S. Dept. Agr. Printed for the use of the committee on Agriculture and Forestry. U.S. Govt. Printing Office, Washington, D.C., 1974.
30. Steinhart, J. S., and Steinhart, C. E. *Science* 184, 307 (1974).
31. Costs of Producing Selected Crops in the United States--1974. Econ. Res. Serv., U.S. Dept. Agr. Printed for the use of the Committee on Agriculture and Forestry. U.S. Govt. Printing Office, Washington, D.C., 1976.
32. Walter, A. S., and Garst, G. D. Fats and Oils Situation, FOS-282, p. 28, Econ. Res. Serv., U.S. Dept. Agr., Washington, D.C., April 1976.
33. Chancellor, W. J., and Goss, J. R., *Science* 192, 213 (1976).
34. Heichel, G. H. *Technol. Rev.*, p. 19, July/August 1974.
35. Agricultural Statistics 1975, p. 137, U.S. Dept. Agr., U.S. Govt. Printing Office, Washington, D.C., 1975.
36. Synthetic Organic Chemicals. United States Production and Sales, 1974. U.S. Intl. Trade Com., ITC Publ. 776, p. 148, U.S. Govt. Printing Office, Washington, D.C., 1976.
37. Bell, F. P., and Nazir, D. J. *Lipids* 11, 216 (1976).
38. Synthetic Organic Chemicals. United States Production and Sales, 1973. U.S. Intl. Trade Com., ITC Publ. 728, p. 159, U.S. Govt. Printing Office, Washington, D.C., 1975.
39. *Chem. Mark. Rep.*, July 28, 1975, p. 9.
40. Van Nieuwenhuyzen, W. *J. Am. Oil Chem. Soc.* 53, 425 (1976).
41. Scocca, P. M. *J. Am. Oil Chem. Soc.* 53, 428 (1976).
42. Pryde, E. H., Gast, L. E., Frankel, E. N., and Carlson, K. D. *Polym.-Plast. Technol. Eng.* 7, 1 (1976).

NEW INDUSTRIAL POTENTIALS FOR CARBOHYDRATES*

F. H. Otey

Northern Regional Research Center
Agricultural Research Service
U.S. Department of Agriculture
Peoria, Illinois 61604

I.   INTRODUCTION

Starch and cellulose are the most abundant polysaccharides
produced in nature. They are now being given special consideration as raw materials for the manufacture of basic organic
chemicals because of increasing prices and decreasing availability of petroleum that is conventionally used to make most
synthetic organic products. Hundreds of new products and applications have been developed in recent years as many companies and governmental agencies have conducted research on
cellulose and starch. Some researchers suggest that these and
other plant products can be used as total replacements for
products made from petroleum.

Before considering polysaccharides for new large-scale
applications, some thought must be given to their chemical
structure and to their present rate of production and consumption as compared to our use of petroleum.

A.   Chemical Structure of Starch and Cellulose

Starch and cellulose are high polymers composed of D-glu-
cose units. Their molecules differ only in the manner in
which the glucose units are joined together.

Cellulose is a linear polysaccharide consisting of 6,000
to 8,000 1,4-linked β-D-glucose units. Because of this 1,4-β-

---

*The mention of firm names or trade products does not
imply that they are endorsed or recommended by the U.S. Department of Agriculture over other firms or similar products not
mentioned.

linkage, these chain molecules can align themselves alongside each other to form linear crystals or microfibrils. The high strength of cellulose fibers may be due to the molecule chain ends being held in lattices by strong hydrogen bonding forces. Also, many of the important properties of cellulose fibers may be due to the molecule chain ends being held in lattices by strong hydrogen bonding forces. Also, many of the important properties of cellulose are attributed to its high resistance to chemical and enzymatic degradation. Cellulose does not gelatinize in hot water, and prolonged heating in strong acid is required to completely hydrolyze the glucosidic bonds of this polymer.

In contrast, most of the common starches contain both a linear polysaccharide (amylose) consisting of 400 to 1,000 1,4-linked α-D-glucose units and a branched molecule (amylopectin) having 10,000 to 40,000 1,4- and 1,6-linked α-D-linkage, the amylose molecules assume a spiral or helical shape having about six glucose units per spiral. The amylopectin molecule consists of D-glucose units linked α-D(1→4) with branches at the C-6 position on an average of once every 26 or more D-glucose units. Most of the common starches contain 17-27% amylose, with the remaining portion being amylopectin. However, there are a number of "waxy" cereal grains in which the starch is entirely of the branched or amylopectin type. Also, because the linear amylose starch offers certain unique film-forming characteristics, breeding studies were conducted which have led to the introduction of new varieties of corn (amylomaize) that produce starch containing up to 70% amylose. One variety of corn, not yet commercialized, is known to produce starch containing 89 to 91% amylose.

Starch, in contrast to cellulose, readily disperses in hot water to form starch-pastes possessing unique viscosity characteristics and good film-forming behavior. Starch is an excellent surface-sizer for paper and textiles because it contributes additional strength to the products. Starch is also readily hydrolyzed in good yields into glucose by either enzyme or chemical treatments.

B.   Current Production and Uses of Cellulose

Cellulose is the most abundant compount in plants. If we can assume that one-third of the reported 40 to 173 billion tons of carbon produced by photosynthesis is converted into cellulose, then the annual world production of cellulose is 30 to 130 billion tons (1). These figures are based on an assumption that considerable carbon fixation occurs in the oceans. Estimates on the annual production of cellulose by land plants range from 22 to 50 billion tons (2,3).

To provide some information on the amount of cellulose used by the United States, we have estimated the cellulose content of major plants consumed annually (Table 1). In 1972,

TABLE 1

U.S. Consumption of Cellulose, 1972

Product	SOURCE		CELLULOSE
	Million tons	Est.%	Million tons
Wood			
Lumber	50.7	40	20.3
Plywood and veneer	12.9	40	5.2
Panels	8.3	40	3.3
Woodpulp for			
Paper and board	46.6	73	34.0
Rayon, acetate, plastics, etc.	1.3	100	1.3
Fuelwood	9.3	40	3.7
Miscellaneous	8.9	40	3.6
Subtotal, Wood	138.0	---	71.4
Others			
Wastepaper for pulp	11.3	73	8.2
Cotton, bagasse, etc., for pulp	0.9	80	0.7
Cotton for apparel, home, and industry	1.7	98	1.7
Subtotal, Others	13.9	---	10.6
Feeds (Dry)			
Hay[a]	93.0	30	27.9
Silage (corn and sorghum)[b]	36.0	30	10.8
Grain (listed in Table 3)	---	2-11	5.5
Subtotal, Feeds	129.0	---	44.2
Total, All	280.9[c]	---	126.2

[a] Based on information gained from references 4 and 5.
[b] Estimated moisture content: silage, 70%; hay, 30%, 1975.
[c] Excludes weight of feed grain.

about 138 million tons of wood products were used as 72 million tons of building materials, 48 million tons of wood pulp, and 18 million tons of fuelwood and miscellaneous products (4).

Most wood contains about 40% cellulose. In the production of
pulp, considerable amounts of lignin and other products are
removed, depending upon the grade of pulp being produced.
Hence, we have arbitrarily estimated the cellulose content of
wood pulp at 73% cellulose.

Another 13.9 million tons of cellulosic products are made
from wastepaper, cotton, and agricultural residues. In 1975,
the United States produced 133 million tons of hay and 120
million tons of corn and sorghum silage (5). We estimated
silage and hay to have 70 and 30% moisture, respectively, and
the dry products of each to have 30% cellulose. Based on
these figures, consumption of cellulose each year is about 126
million tons. Moreover, other cellulosic products not listed,
such as food and pasture, are used.

In addition to the products listed in Table 1, about 50
million tons of cellulose is present in various unused agricultural and forest product residues (Table 2). Even though
75% of all wood residue from sawmills and other primary wood

TABLE 2

*Unused Agricultural and Forest Product
Residues in the United States*

Product	SOURCE Million tons	CELLULOSE Est.%	CELLULOSE Million tons
Wood residues, 1970[a]			
Primary manufacturing plants	13		
Logging	21		
Limbs, dead or rotten trees	21		
Total, Wood	55	40	22
Nonwood plant fibers, 1972[b]			
Sugar cane bagasse	1.8		
Straw	74.3		
Cotton staple fiber	3.3		
Second cut cotton linters	0.3		
Total, Nonwood	79.7	35	28
Total	134.7		50

[a]Calculated from information given in reference 4.
[b]Calculated from information given in reference 6.

processing plants is used, there remains about 4.2 billion
cubic feet, or an estimated 55 million tons of unused wood
residue from logging and various milling operations (4). In
addition, Atchison reports that about 80 million tons of col-
lectable nonwood agricultural residues are left unused as
sugar cane, straw, and cotton products (6). There are a number of cellulosic waste products not listed in Table 2. One
report estimates that all wood waste, industrial and consumer
waste, farm waste, and other sources produce 700 million tons
annually of cellulosic products that contain 224 million tons
of cellulose (7). We now produce only 0.5 million tons of
paper pulp from agricultural residues. Any substantial
increase in the use of these wood and agricultural residues
will require a significant breakthrough in collection techniques. Also, any thought of using major amounts of agricultural residues must be balanced against benefits gained from
allowing these residues to decay as soil-conditioning agents.

C.   Current Production and Uses of Starch

While cellulose is the most abundant compound in plants,
starch is the most abundant reserve polysaccharide. Vigorous
starch synthesis takes place in the grains of cereal plants.
Seven major U.S. starch-producing crops are listed in Table 3
along with the percentages of protein and total amount of

TABLE 3

Nonfiber Carbohydrates Produced by Seven U.S. Crops, 1975

Crop	Yield[a] Million bu.	Yield[a] Billion lb.	Protein (8)	(8)	Nonfiber carbohydrate[b] Billion lb.	Nonfiber carbohydrate[b] Million tons
Corn	5,767	322.9	9.7	71.1	229.6	114.8
Wheat	2,134	128.0	13.2	69.9	89.5	44.8
Sorghum	758	42.5	11.3	71.3	30.3	15.2
Oats	657	21.0	12.0	60.2	12.6	6.3
Barley	383	18.0	11.8	68.0	12.2	6.1
Rice	---	12.8	7.9	64.9	8.3	4.1
Potatoes	---	31.6	---	16.6	5.2	2.6
TOTAL		576.8			387.7	193.9

[a] Calculated from data in reference 5.
[b] Values are 10% or more too high because moisture level was
not considered.

nonfiber carbohydrate present in their produce (5,8). Virtually all of this carbohydrate is starch. Corn produces about
59% of the total starch and wheat 23%. About 66% of all corn
is used for feed and 9% goes for food, industry, and seed.
The remaining is either exported or stored (9).

We isolate only 2 to 3% of the available starch from
these crops. Most of the starch isolated in the U.S. is obtained from corn by a wet milling process. About 6% of the
total 1974 corn crop was processed for starch. Although exact
figures were not available to us on production and uses of
starch, Table 4 provides a reasonable estimate of these figures

TABLE 4

U.S. Corn Starch Consumption, 1974-1975

Uses	Billion pounds, dry basis[a]	
	1974	1975
Industrial corn starch	3.42	2.95
Dextrins	0.13	0.10
Subtotal	3.55	3.05
Sweeteners		
Regular corn syrup	3.60	3.45
High fructose	0.80	1.45
Dextrose	1.25	1.15
Miscellaneous refinery products	0.15	0.15
Subtotal, Starch for Sweeteners[b]	5.22	5.58
Total starch consumed	8.77	8.63

[a]Estimates only; sweetener values from reference 10; others
from private conversations. Starch and dextrin values reduced to account for an estimated 10% moisture level.
[b]Total sweetener weight multiplied by 0.9 to correct for 1
mole of added H_2O.

for 1974 and 1975. We chose to report starch on a dry basis
and to include the nonsweetener food starch with the industrial starch. The starch figures have been multiplied by 0.9
to correct for an estimated 10% moisture that is usually present in starch. Also, to obtain the estimated total starch
used to make sweeteners, we multiplied the total sweetener
value by 0.9 to account for 1 mole of water added during
hydrolysis of the starch.

In 1974, about 60% of all corn starch was converted into

low-molecular-weight sugars; in 1975, this figure increased to 65%. The major change in sugar production was due to increased production of high-fructose syrups (10). Because of the unstable market for sucrose, the soft drink industry has started using significant amounts of high-fructose corn syrup sweetener. In 1974, the price of cane sugar rose 482% within 12 months. Although sucrose prices dropped in 1975, long-term projections are still showing cane and beet sugar in the 18 to 20¢ per pound range (11). The U.S. production capacity for high-fructose syrup is expected to reach 4.8 billion pounds by 1978 (12). Kolodny (10) expects the production of dextrose and regular corn syrup to level off at 1.0 and 3.5 billion pounds respectively during the period 1976 to 1980. During this same period, he believes that the yearly production of high-fructose corn syrup will reach 4 billion pounds. Hence, by 1980, we may be using about 2.6 billion pounds more starch for sweeteners than we are reporting for 1974, which represents the fastest growing market for starch that the industry has ever experienced.

Although the consumption of starch for sweeteners in 1975 increased by about 360 million pounds over the 1974 figure, other commercial applications of starch decreased by about 500 million pounds. Most of the lower industrial starch applications in 1975 were caused by reduction in paper and textile production. Russell (13) reports that we have just about recovered from the dip in paper production that bottomed out in 1975.

In 1974, about 40% of all isolated corn starch, or 1% of the total available starch, was used for nonsweetener industrial applications (Table 5). Although starch and cellulose

TABLE 5

U.S. Industrial Starch Applications, 1974

Uses	Amounts, dry basis[a]	
	%	Million lb.
Paper	60	2,130
Textiles	10	355
Food	14[b]	497
Miscellaneous	16	568
TOTAL	100	3,550

[a] Estimates gained from private conversations.
[b] Includes amount used to make beer.

are often produced by nature in the same plant, they are
always isolated from each other for industrial applications.
Yet, the largest single industrial application for both of the
carbohydrates is in paper production, where they are recombined into a single product. Of the 3.55 billion pounds, dry
basis, of industrial starches used in 1974, 60% or 2.13 billion pounds was used in paper, including paperboard and corru-
gated products. Hill (11) suggests that the United States is
now using 2.8 (probably 2.5 dry basis) billion pounds of
starch in these paper products and that between 1976 and 1980
this figure will increase by 25%. Hence, by 1980, we may be
using 1 billion pounds more starch for paper production than
is reported for 1974.

About 10% of industrial starch goes to the textile industry, and another 14% is used by food and brewing industries.

At least 40% of all industrial starch is modified, before
being marketed, by such treatments as partial depolymeriza-
tions, oxidation, and reaction with propylene oxide, ethylene
oxide, and various amines. Generally, these modifications are
carried out to lower the viscosity and to improve the clarity
and stability of aqueous starch dispersions. In other instances these modifications are needed to give the starch
greater affinity for natural or synthetic fibers (14).

D.   Comparison of Starch, Cellulose, and Petroleum Uses

Despite the large amounts of starch and cellulose products used for lumber, paper, clothing, feed, and food, their
total is substantially lower than the annual consumption of
petroleum (Table 6). We use about 13% as much cellulose and

TABLE 6

Summary of Cellulose, Starch, and Petroleum Products

Source	Total, Million tons	Carbohydrates, Million tons
Cellulosic Products, Total Table 1	281	126 Cellulose
Unused Cellulosic Residues, Table 2	135	50 Cellulose
Subtotal, Cellulosic	416	176
Starch Crops, Table 3	288	194 Starch
TOTAL, Starch and Cellulose	704	370
TOTAL, Petroleum, 1974	956	---

20% as much starch as we do petroleum. The combined weight of
all residues listed in Table 2, and all grains and potatoes
listed in Table 3 is only 74% of the total weight of petroleum
consumed each year.

Since natural plant products are already consumed at a
rate nearly equivalent to their production, any attempt to use
them as a major replacement for petroleum must depend on
sources other than those discussed. Also, consideration must
be given to the increasing needs predicted for these carbohydrates in their present applications.

However, we do believe that sufficient amounts of cellulose and starch can be made available to significantly reduce
the demand on petroleum for chemical feedstocks. About 7 to
8% of all liquid petroleum supplies are now used for chemical
feedstocks and, unless other supplies are made available, 20%
of the petroleum may be needed for feedstocks by 1985 (15). A
significant amount of cellulose, suitable for chemical processing, should be available from certain wood and agricultural
residues that do not have fiber qualities suitable for pulp
(16). Starch may become increasingly available from grain
processors upgrading the protein content of certain cereal
products by removing part of the starch. We have already received inquiries on how wheat and oat starches might find new
industrial applications.

E.   Potential New Uses for Cellulose

Currently, about 2 to 4 billion pounds of refined cellulose is used to produce synthetic products such as cellulose
acetate, rayon, and plastics. This figure represents only
about 1 to 2% of the total cellulose consumed in wood, cotton,
and feed products. As the price of petrochemicals has increased, cellulose has achieved a more favorable economic
position in synthetic applications, particularly in plastics.

Although cellulosic materials have been used for many
years as reinforcing agents in thermosetting plastics, they
have found very limited application in thermoplastics. Quick
reviewed a number of recent studies which suggest that cellulose can be used to improve the cost, mold shrinkage, flexibility, reinforcement, and biodegradability of certain thermoplastics (17).

Goldstein (16) reviewed the economics of converting cellulose and other wood products into chemicals and fuels that
are normally obtained from petroleum. Apparently a process
has been developed by the Bureau of Mines for converting wood
into liquid fuel at a cost of $15.64 per 420-pound barrel,
assuming the cost of wood at $20 per ton. This is about $3 to
$4 more per barrel than the cost of imported oil. Also, a
process has been developed for converting cellulose into

ethylene and butadiene in total weight yields of about 16 and
11%, respectively. Of the 36 billion pounds of synthetic
polymers used each year, 47% are derivable from ethylene and
12% from butadiene.

In another approach, Klass believes that the U.S. oil and
gas requirements can be met with biomass conversion of plant
growth derived from intensive cultivation of about 6% of the
land area of the conterminous 48 states. The conversion process depends upon high-temperature oxidation or biological
conversion of the total plants into a variety of hydrocarbons
plus CO, H_2, and NH_3 (18).

F.   Potential New Uses for Starch

In addition to the rapidly expanding market for starch in
sweeteners and paper, a number of starch and starch-derived
products have been developed which have industrial potential.
One of the most promising of these is in plastics. In 1974,
28.5 billion pounds of plastics were produced in the United
States and, if raw materials are available, this figure may
double by the mid 1980's (19). Although numerous developments
have been reported on how starch can be used to make plastics,
these have failed to achieve large-scale market reality,
largely because lower cost plastics could be made from petrochemicals. However, increasing prices and demands for plastic
resins, coupled with public desires for biodegradable and
flame-resistant plastics, are forcing industry to consider
starch as a raw material for plastics. Starch is highly biodegradable and flame-retardants are easily attached to certain
starch derivatives. Some of the more promising products in
which starch may find new application are listed below.

1.   *Polyurethane foams*

One type of plastic that affords an excellent opportunity
for starch application is rigid urethane foam. Currently the
U.S. produces about 385 million pounds of rigid urethane foam,
and by 1980 this annual production figure could reach one billion pounds. These foams gained rapid acceptance because they
are easy to apply, give a high strength-to-weight ratio, and
have about twice the insulating value of any other commercially available material. Such foams contribute to low-cost
housing and conserve heating and cooling energy.

The chemistry of urethane foams involves crosslinking
polyethers with isocyanates in the presence of catalysts,
blowing agents, and surfactants. Polyethers used in these
formulations can be prepared from starch-derived products. In
the early development of polyethers, sorbitol, obtained by
hydrogenating glucose, served as a raw material. Currently,
only 1 to 2 million pounds of sorbitol go into making poly-

ethers for urethane foam. Possible factors influencing this
restricted use of sorbitol are cost and the fact that the
flexible straight chain of sorbitol yields foams with slightly
poorer flame resistance than do cyclic glycosides.

Pentaerythritol, derived from petroleum, and sucrose became the leading raw materials for polyether manufacturing.
Recent fluctuations in price and availability of these materials have caused polyether manufacturers to reconsider starch-
derived products as a basic polyol.

In addition to possible revived interest in sorbitol and
methyl glucoside, an experimental polyol obtained from the
reaction of ethylene glycol with starch may find large-scale
application in the polyether market (20). These polyols are
prepared by heating a mixture of starch, ethylene glycol, and
sulfuric acid catalyst for 1 hr. Although the product of this
reaction is a mixture of polyols, it need not be purified for
use in polyether production. The crude polyol mixture is
treated with about 6 moles of propylene oxide per mole of
polyol to give the polyether. Although these polyethers received considerable industrial attention in the mid 1960's,
when we first developed them, they are now being evaluated
seriously for industrial production. In 1975, two companies
were granted licenses to practice the process.

To provide some information on the potential market for
starch as a raw material for urethane foam, theoretical amounts
were calculated of materials required to produce polyethers
from three starch-derived polyols (Table 7). These figures

TABLE 7

Potential Market for Starch in Urethane Foams

| Polyol | Materials to make 100 lb. Polyether with Hydroxyl Number of 450 lb. | | | Starch/ billion lb. foam, MM lb. |
	Propylene oxide	Others	Starch	
Sorbitol	75.7	$H_2 + H_2O$ 2.6	21.7	82
Methyl glucoside	61.0	CH_3OH 6.4	32.5	122
Glycol glycosides	63.0	Glycol 8.6	28.0	105

show that the projected 1980 market of one billion pounds of
foam could require more than 100 million pounds of starch.

Although urethane foams without flame resistance are used
as insulation in many places, such as refrigerator walls, they

cannot be used in building. Flame-resistance is usually built
into foams by adding halogens, phosphorus, and antimony compounds. However, any such foams must be sandwiched between
fireproof materials to meet housing codes. We have developed
flame-retardant polyethers by the stepwise reaction of glucose
with allyl alcohol, propylene oxide, and halogenating compounds (21). In the first step, allyl alcohol readily reacts
with glucose to form a mixture of glucosides having an unsaturated site for addition of flame retardants. This mixture is
combined with 4 to 5 moles of propylene oxide and then with a
halogenating agent such as bromine or carbon tetrachloride.
Data from laboratory studies suggest that centrally locating a
halogen in the rigid cyclic structure of a glucoside moiety
greatly improves its efficiency as a flame retardant.

2.   *Biodegradable plastics*

To alleviate disposal problems, films and rigid plastics
that will biodegrade are now sought for many applications.
One multimillion-pound-per-year application for plastic film
where a biodegradable film is urgently needed is agriculture
mulch. Farmers find that they can greatly improve the yield
and quality of vegetable and fruit crops by covering prepared
rows with plastic films and then transplanting crops through
holes punched in the films. Polyethylene films are commonly
used for mulch, but since they do not degrade between growing
seasons they must be removed from the field at a cost of $40
to $100 per acre. Hence, farmers have expressed a desire for
a mulch film that will degrade between growing seasons to
obviate this troublesome and expensive removal procedure.
Recent studies on the incorporation of starch into plastic
films indicate that prospects for producing a satisfactory
biodegradable film are good. In one approach, a water disper-
sion of starch, polyvinyl alcohol, and plasticizer is cast
onto a plate, oven dried, and finally made into water-resis-
tant film with a coating of Saran or polyvinyl chloride (22).
Although these films showed acceptable properties during artificial weathering, additional study is needed to improve processing techniques and properties of these films.
Since early in 1975, one company has been using this
starch polyvinyl alcohol technology, with minor modifications,
to produce a water-soluble laundry bag. The modification involves using slightly derivatized starch, in place of pearl
starch, to impart water solubility. The bags are used in
hospitals to keep the laundry staff from coming in contact
with soiled or contaminated clothing. The unopened bags are
placed into washing machines where they readily dissolve.
These plastic bags may also find application as packaging for
agricultural chemicals.

Starch can also be incorporated as a filler into conventional resins such as polyethylene and polyvinyl chloride (PVC). Coloroll Ltd., England, is now producing a polyethylene bag that is reportedly biodegradable because it contains 7 to 10% starch filler. We developed techniques for incorporating large amounts of starch filler into PVC films and rigid plastics (23). The PVC-starch plastics are readily attacked by soil microorganisms, which suggests that they will biodegrade upon disposal. Up to 40% starch was added to the plastics formulations without significantly affecting tensile strength and clarity of the plastics; however, starch addition greatly reduced the percentage elongation. These starch-PVC plastics may find applications in packaging, food trays, plates, and other items that are used once or twice and then discarded.

3. *Plastic-Related Applications*

In addition to the developmental starch products mentioned for plastics, a number of others hold considerable promise in plastic-related areas.

The discovery that starch can serve as the backbone for grafting synthetic polymers has led to a number of interesting starch copolymers. Notable among them is a hydrolyzed starch-polyacrylonitrile which can absorb up to 1,400 times its weight of distilled water. This polymer, called Super Slurper, is potentially useful for controlling soil moisture around seeds and seedlings, in surgical sponges, in diapers, and the like. Super Slurper is now being produced on a limited scale, and steps are being taken to implement large-scale production (24).

Starch xanthide is a good substitute for low and medium grades of carbon black used to reinforce rubber (25). Current domestic use of carbon blacks in rubber is about 3 billion pounds annually. In a related study, an economically feasible process was developed for making powdered rubber from a variety of rubber latices (26). Conventionally, these latices are precipitated into large rubber slabs that require exorbitant milling energy to blend them with fillers, stabilizers, and curing agents. In the new process, techniques were developed for producing a coprecipitate, containing 3 to 5% starch xanthide and 95 to 97% rubber, from a latex-starch xanthate mixture. Preliminary calculations, based on an estimated 50% market penetration, suggest that the United States could save 2.5 billion kilowatt hours annually by using the new process.

II. REFERENCES

1. Robinowitch, E., and Govindjee. "Photosynthesis," John Wiley and Son, p. 12. 1969. New York.
2. Bonner, J., and Varner, J. E., 1965. "Plant Biochemistry,"

p. 307. 1965. Academic Press, New York.

3. Goldstein, I. S. Abstr. Paper 4, Div. Cellulose, Paper and Textile Chem., Macromolecular Scretariat, 171st Amer. Chem. Soc. Meeting, New York, April 5-9, 1976.

4. U.S. Dep. Agric., Forest Service, the Outlook for Timber in the U.S., Forest Resource Rept. No. 20, July 1974.

5. U.S. Dep. Agric., Statistical Reporting Service, Crop Production, 1975 Annual Summary, Washington, D.C., January 15, 1976.

6. Atchison, J. E. *Pap. Trade J.* 157(2), 33 (1973).

7. Anon. *Chem. Eng. News* 54, 12 (1976).

8. Matz, S. A. "Cereal Science," p. 229. 1969. Avi Publishing Co. Ind., Westport, Conn.

9. Anon. 1975 Corn Annual p. 13. 1975. Corn Refiners Assoc. Inc., Washington, D.C.

10. Anon. *Chem. Eng. News* 54(17), 13 (1976).

11. Anon. *Pap. Age* 12(3), 19 (1976).

12. Anon. *Milling Baking News* 54(44), 34 (1975).

13. Russell, C. R. Corn Annual, p. 26. 1976. Corn Refiners Assoc. Inc., Washington, D.C.

14. Russell, C. R., *Symp. Proc. Am. Assoc. Cereal Chem.*, p. 262, St. Louis, Mo., November 4-8, 1973.

15. Anon. *Chem. Eng. News* 54(14), 6 (1976).

16. Goldstein, I. S. *Science* 189, 847 (1975).

17. Quick, J.R., Symp. Pap. Div. Chem Marketing Econ., 169th Amer. Chem. Soc. Meeting, p. 56. Philadelphia, Pa., April 7-9, 1975.

18. Anon. *Chem. Eng. News* 54(8), 24 (1976).

19. Anon. *Mod. Plast.* 53(1), 39 (1976).

20. Otey, F. H., Zagoren, B. L., and Mehltretter, C. L., *Ind. Eng. Chem., Prod. Res. Dev.* 2, 256 (1963).

21. Otey, F. H., Westhoff, R. P., and Mehltretter, C. L., *J. Cell. Plast.* 8, 156 (1972).

22. Otey, F. H., Mark, A. M., Mehltretter, C. L., and Russell, C. R. *Ind. Eng. Chem., Prod. Res. Dev.* 13, 90 (1974).

23. Westhoff, R. P., Otey, F. H., Mehltretter, C. L., and Russell, C. R. *Ind. Eng. Chem., Prod. Res. Dev.* 13, 123 (1974).

24. Weaver, M. O., Fanta, G. F., and Doane, W. M., Proc. Tech. Symp. Nonwoven Product Technol., Int. Nonwoven Disposables Assoc., p. 169, Washington, D.C., March 1974.

25. Buchanan, R. A., Kwolek, W. F., Katz, H. C., and Russell, C. R. *Staerke* 23, 350 (1971).

26. Abbott, T. P., James, C., Doane, W. M., and Russell, C. R. *Rubber World* 169, 40 (1974).

THE CURRENT IMPORTANCE OF PLANTS AS A SOURCE OF DRUGS

Norman R. Farnsworth

Department of Pharmacognosy and Pharmacology
College of Pharmacy
University of Illinois at the Medical Center
Chicago, Illinois 60612

I.   INTRODUCTION

The necessity and importance of plants to produce foods for man and animal is rarely disputed, albeit often taken for granted. On the other hand, it is generally presumed by most that drugs, and especially prescription drugs, are primarily of synthetic origin. Using data from National prescription surveys, and other available data, it can be shown that at least 25 per cent of prescriptions dispensed in the United States, contain one or more active ingredients that are still extracted from plants. This market is estimated at about $3 billion, at the consumer level, in 1973.

Although the main basis of this presentation will be concerned with the importance of plants as drugs in the United States, one cannot neglect to at least briefly point out that in other parts of the world, plant drugs are even more important than in our own country.

II.   IMPORTANCE OF PLANT DRUGS IN OTHER COUNTRIES

In 1974, this author was privileged to visit the People's Republic of China as one of a 12-member "Herbal Pharmacology Delegation", for 26 days. We visited research institutes, Traditional hospitals, Western hospitals, Traditional drug manufacturing plants, community pharmacies, hospital pharmacies, and other institutions where it was possible to see the great dependence that 850 million Chinese people have on plants as a major contributing factor to an effective health-care system. That the medical system in the People's Republic of China is difficult to evaluate becuase of the combined use of Traditional Chinese medicine, i.e. use of herbs and acupunc-

ture, as well as Western-type medicine, cannot be denied.
However, it is evident that plant drugs are widely used in the
health-care system, and since our delegation concluded that
the people in China were receiving at least equivalent health
care as we in the United States, lends some credence to the
possibility that plant drugs are a contributing factor to this
success. The report of our delegation has been published (1),
and can be consulted for the results of our analysis of the
role of plants in the health-care system in China.

Further, there is current interest to find external
sources of drug plants that are needed by an estimated 1 billion users in the Orient, since the demand has exceeded current supplies (2).

According to Chikov (3), there is a distinct reversal in
the trend to use synthetic drugs in the U.S.S.R. It is claimed that the annual requirement for medicinal plants in the
U.S.S.R. is 40,000 tons, approximately equally divided as being obtained from cultivated sources as from wild-growing
plants (3). It has been estimated that it will be necessary
to double this production within the next 5-10 years.

At a recent Symposium on "Traditional Medicine Evolution
in the Contemporary Society", held in Mexico, and attended by
the author, the dependence on plants as sources of drugs in
such countries as Nigeria, Ghana, India, Madagascar, Mexico,
Guatemala, People's Republic of China and Egypt, was emphasized. It was claimed that the African continent is dependent
on as much as 95 per cent of it's drug requirements from
plants.

This "Green Drug Revolution" has appeared to make an impact here in the United States, where the question of preservation of germ plasm for drug plants is now being studied by
the National Research Council.

III. IMPORTANCE OF DRUG PLANTS IN THE UNITED STATES

One need not be a scientist to realize the mass prolifer-
ation of herb outlets here in the United States. The herbal
industry is growing by leaps and bounds, and apparently has
now begun to draw the attention of the Food and Drug Adminis-
tration, who in March of 1977, initiated regulatory policy on
a large number of plants, including the following: *Vinca
minor*, *Vinca major*, *Ipomoea purpurea*, *Datura stramonium*,
Atropa belladonna, *Viscum album*, *Artemisia absinthium*, and
others. It will be interesting to see how the FDA intends to
regulate so many plants that are common garden ornamentals,
ground covers, and plants that grow everywhere as weeds. In
any event, the herb trade industry is growing, and is important to keep in mind with regard to any discussion on the importance of plants as drugs. At the present time, it is im-

possible to attach any accurate significance to this industry
in terms of dollars or other meaningful experssion.

In addition, one cannot rule out the impast on the health-
care system in the United States for such over-the-counter
drugs as plantago seed, cascara bark, senna leaves and pods,
scopolamine, castor oil, etc., which are infrequently prescribed as drugs, but nevertheless, exert an important influence on the drug market.

The major thrust of this presentation is, however, to
asses the relative importance of plants as sources of prescription drugs.

To the best of my knowledge, data are not available outside the U.S.A. that allow one to calculate the actual number
of prescriptions dispensed to patients that contain plant-
derived drugs, nor the monetary value of such prescriptions.
However, we can now document rather well that in the United
States in the year 1973, the American public paid about $3
billion for prescription drugs that are still extracted from
higher plants.

Recent data (4) claim that domestic sales of ethical
drugs (at the manufacturer's level) in the U.S.A. totaled $6.3
billion in 1974 for human dosage forms, and that world-wide
sales of combined veterinary and human dosage forms totaled
$11.3 billion in the same year. One can probably double these
industry figures to estimate the cost of human and/or veterinary drugs to the consumer.

We have analyzed the National Prescription Audit (NPA)
data (5) in the U.S.A., which includes total new and refilled
prescription sales for community pharmacies in the United
States. Of the 1.532 billion prescriptions (5,6) dispensed
during 1973, 25.2% contained one or more active constituents
obtained from higher plants (seed plants). If one considers
that in 1973, the average prescription price to the consumer
was $4.13 (6), then total prescription sales in community
pharmacies for drugs from higher plants for that year amounted
to about $1.59 billion. further, microbial products (anti-
biotics, ergot alkaloids, immunizing biologicals, etc.) accounted for about 2.7% of the total.

In order to determine whether or not 1973 was an atypical
year, a computerized analysis was carried out on the American
prescription market from NPA data each year for the period
1959 through 1973. Although the total number of prescriptions
increased dramatically over this 15-year period, the percentage of natural-product prescriptions remained rather constant
(Table 1), indicating perhaps two major points: (a) that
natural products represent an extremely stable market in the
United States, and (b) that, because of this stability, it can
be safely assumed that the drugs represented in the survey are
heavily relied on (prescribed) by physicians.

TABLE 1

Comparison of Natural-Product Containing
Prescriptions Dispensed in Community Pharmacies (1959 and 1973)

Year	Higher Plants	Microbes	Animals	Total
1959	25.5%	21.4%	2.3%	49.2%
1973	25.2%	13.3%	2.7%	41.2%

While it is true that the total percentage of prescriptions containing natural products decreased from 49.2% in 1959 to 41.2% in 1973, it is clear from Table 1 that the drop was attributed solely to a decreased use of microbial products, chiefly antibiotics. Thus, it can be stated that over the period 1959-1973, drugs from higher plants did not increase or decrease in frequency of use in the American prescription market. This is of interest because no new drugs from higher plants were introduced during the same span of time. It is known that industry research and development investment for higher drug plant research during this same period of time decreased substantially. During the period 1959-1973, it is known that in the U.S.A., research programs in the pharmaceutical industry relating to the search for new drugs from higher plants were either phased out or reduced at Ciba, Smith Kline and French, Riker, G. D. Searle, and Eli Lilly and Co., and perhaps at other pharmaceutical companies as well.

National Prescription Audit figures for 1973 (6) indicate that 1.532 billion new and refilled prescriptions were dispensed from community pharmacies in the United States. At an average cost to the consumer of $4.13 per prescription (6), one can calculate a dollar value of $6.327 billion for the market in 1973. Thus, if a predicted 25.2% of these prescriptions contained active principles of higher plant origin, the dollar cost to the consumer in 1973 would be estimated at $1.594 billion.

Now, how does one obtain the figure of $3 billion as the current value of higher plant medicinals in the U.S.A.? It can be estimated that somewhat less than the dollar volume representing the community pharmacy prescription market may be added to the $1.594 billion prescription market to account for the volue of drugs dispensed in hospitals, government agencies, and the like. Thus, it seems logical and convenient to consider $3 billion as the annual value of drugs at the consumer level that are obtained from higher plants.

To illustrate the importance of many higher plant drugs,

the 12 most commonly encountered pure compounds, derived from
higher plants and tabulated from the 1973 NPA prescription
data, are presented in Table 2.

TABLE 2

Most Commonly-Encountered Pure Compounds
From Higher Plants Used as Drugs in 1973 in the U.S.A.

Active plant principle	Total number of Rxs[a]	Per cent of Total Rxs
Steroids (95% from diosgenin)	225,050,000	14.69
Codeine	31,099,000	2.03
Atropine	22,980,000	1.50
Reserpine	22,214,000	1.45
Pseudoephedrine[b]	13,788,000	0.90
Ephedrine[b]	11,796,000	0.77
Hyoscyamine	11,490,000	0.75
Digoxin	11,184,000	0.73
Scopolamine	10,111,000	0.66
Digitoxin	5,056,000	0.33
Pilocarpine	3,983,000	0.26
Quinidine	2,758,000	0.18

[a]Total number of Rxs in 1973 was 1.532 billion
[b]Produced commercially by synthesis, all others by extraction
from plants

Another interesting note is that in 1973, a total of 76
different chemical compounds of known structure, derived from
higher plants, were represented in the prescriptions analyzed.
Further, the assumption by many people is that most, if not
all, of the higher plant-derived drugs of known structure are
now produced commercially by synthesis. Nothing could be
further from the truth. Of the 76 individual drugs just indicated, only seven are commercially produced by synthesis, eme-
tine, caffeine, theobromine, theophylline, pseudoephedrine,
ephedrine, and papaverine. This is not to imply that most of
the naturally-occurring drugs have not been synthesized;
indeed they have. However, practical industrial synthesis for
such important drugs as morphine, codeine, atropine, digoxin,
digitoxin, etc. are not available. The alkaloid, reserpine,
for example, can be commercially extracted from natural sources
for about $0.75/gm. whereas a multi-step and difficult syn-
thesis is available that yields reserpine at about $1.25/gm.
It should be obvious which of the two sources is used to pro-

duce this pharmaceutical.

Even more interesting information can be derived from the 1973 survey data. For example, 99 different crude plant drugs, or types of extracts from crude plant drugs, were found to be present in the prescriptions analyzed, involving about 38,300,000 prescriptions in 1973 (2.5% of the total). Those found in the greater number of prescriptions are listed in Table 3.

TABLE 3

Most Commonly Encountered Higher Plant Extracts Used in Prescriptions in 1973[a]

Crude Botanical or Extract	Total number of Rxs	Per cent of Total Rxs[b]
Belladonna *(Atropa belladonna)*	10,418,000	0.45%
Ipecac *(Cephaelis ipecacuanha)*	7,047,000	0.46%
Opium *(Papaver somniferum)*	6,894,000	0.45%
Rauwolfia *(Rauvolfia serpentina)*	5,822,000	0.38%
Cascara *(Rhamnus purshiana)*	2,451,000	0.16%
Digitalis *(Digitalis purpurea)*	2,451,000	0.16%
Citrus Biflavonoids *(Citrus* spp.*)*	1,379,000	0.09%
Veratrum *(Veratrum viride)*	1,072,000	0.07%

[a]Compounded prescriptions represented less than 2.0% of total prescriptions (6) and were excluded from the survey data that were compiled and analyzed. The drugs indicated above were in standard dosage forms and not in multi-component, extemporaneously prepared prescriptions.
[b]Total Rx volume in 1973 was 1.532 billion prescriptions.

One only needs to open the pages of any standard textbook of pharmacology to be impressed by the fact that virtually every pharmacological class of drug includes a natural product prototype that exhibits the classical effects of the pharmacological category in question; most of them are plant-derived (see Table 4).

IV. NON-DRUG BIOLOGICALLY ACTIVE PLANT SUBSTANCES

Natural drug products, many of which have been derived from higher plants, play an important role as useful investigative tools in pharmacological studies. Some such compounds are included in Table 4. Others are mescaline and LSD-derivatives in the study of psychiatric disorders; various toxins,

TABLE 4

Typical Plant Principles Used to Illustrate Pharmacological
Principles in Standard Textbooks

Type of Pharmacological Action	Type of Compound	Name of Compound
Centrally Acting Skeletal Muscle Relaxant	Alkaloid	Bulbocapnine
Analgesic	Alkaloid	Morphine, Codeine
Smooth Muscle Relaxant	Alkaloid	Papaverine
Anti-Gout	Alkaloid	Colchicine
CNS Stimulant	Monoterpene	Camphor
	Sesquiterpene	Picrotoxin
	Alkaloid	Strychnine, Caffeine, Theobromine, Theophylline
Local Anesthetic	Alkaloid	Cocaine
Parasympatholytic	Alkaloid	Atropine, Scopolamine
Parasympathomimetic	Alkaloid	Pilocarpine, Physostigmine
Peripherally Acting Skeletal Muscle Relaxant	Alkaloid	d-Tubocurarine
Sympathomimetic	Alkaloid	Ephedrine
Ganglionic Blocker	Alkaloid	Nicotine, Lobeline
Cardiotonic	Cardiac glycoside	Digitoxin, Digoxin
Antiarrhythmic	Alkaloid	Quinidine
Uterine Stimulant	Alkaloid	Sparteine, Ergot Alkaloids
Antihypertensive	Alkaloid	Reserpine, *Veratrum* Alkaloids
Psychotropic	Alkaloid	Reserpine
Cathartic	Anthraquinone	Anthraquinone glycosides
	Mucilages	Psyllium, Agar
	Fixed Oil	Castor Oil
Antimalarial	Alkaloid	Quinine
Antiamebic	Alkaloid	Emetine

e.g. tetrodotoxin, in the study of nerve transmission; cyclo-
pamine in the study of teratogenesis; phalloidin for induction
of hepatoxicity; and phorbol myristate acetate as a standard
co-carcinogen in the investigation of potential carcinogens
and co-carcinogens.

Other useful applications of plant-derived chemicals can
be cited *e.g.* bixin as a coloring agent for foods; nordihydro-
guaiaretic acid as an antioxidant in lard; and essential oils
and their derived terpenes as perfunes and flavoring agents.
The economic value of these materials is difficult to estimate,
but surely it must be in the billion dollar category on a
world-wide basis.

A number of researchers feel that the major purpose for
finding in plants new structures having biological activity is
to provide templates for the synthesis of analogues and/or
derivatives which will have equivalent or better activity than
the parent molecule. This may indeed by an admirable purpose,
and from a practical point-of-view, it may be advantageous
with regard to patent protection. However, history shows that
it is an exceptionally rare instance when a naturally occurring
chemical compound that has found utility as a drug in man, will
yield a derivative on structure modification that exceeds the
value of the parent compound in drug efficacy.

This also does not discount the value of such model compounds as cocaine, yielding information that led chemists to
produce related local anesthetics such as procaine and its
congeners, nor the value of the large number of synthetic anti-
cholinergic drugs that were designed from the tropane nucleus
and which have their own specific advantages.

Finally, the value of plant-derived chemical compounds as
building blocks for semi-synthetic derivatives cannot be under-
estimated. The classical example is the use of diosgenin as
the primary starting material for the synthesis of the majority of steroidal hormones currently used in medicine.

V.   CURRENT LEVEL OF WORLD-WIDE RESEARCH ON PLANT-DERIVED
     DRUGS

No one to date has compiled accurate figures for a given
year, concerning the number of novel compounds *vs.* the number
of compounds of known structure reported from plants, fungi,
and animals. Since in our laboratory we now are computerizing
the world literature on all aspects of natural products, we
are able to present some interesting data that we consider to
be quite accurate. Before presenting the data, however, an
explanation will be given regarding the extent of the literature covered. Our sources are *Chemical Abstracts*, *Biological
Abstracts*, and current issues of 140 journals which we have
found by experience to contain the majority of new information

concerning natural products. From these sources, we computerize all reports concerning new biological activities for novel
or known compounds from natural sources, and all reports concerning biological activities reported for extracts prepared
from organisms. We computerize, rather completely, reports in
which a secondary metabolite has been isolated from, or detec-
ted in, any living organism.

The data to be cited are from about 10,000 articles, the
contents of which we computerized in the year 1975. No 1976
literature will be discussed; however, a large segment of 1974,
and even some earlier, data are represented, in addition to
data published in 1975, because of the fact that many references in abstracting services tend to be less than current.

The figures to be cited now, represent number of compounds
of already known and of novel structure, reported isolated
from various groups of organisms during the 1975 literature as
defined above. Compounds identified by chromatographic procedures, without being actually isolated, are not included.

TABLE 5

Numbers of Compounds of Known Structure Isolated
from Living Organisms as Determined from the 1975 Literature

Organism Group	New	Known
Monocots	97	277
Dicots	1,504	2,579
Gymnosperms	49	221
Pteridophytes	29	90
Bryophytes	17	32
Lichens	14	44
Fungi and Bacteria	479	523
Marine Organisms	210	199
Totals	2,399	3,965

As can be seen in Table 5, about 1,650 new structures
were reported as isolated from higher plants in the 12-month
audit period, as well as about 3,077 previously known compounds. If one considers compounds isolated from all types of
organisms, then about 2,400 new structures and 4,000 previously isolated compounds can be counted.

To say that these compounds have little commercial importance would not be true; in 1975 more than 400 patents were

TABLE 6

Summary of Classes of Plant Products Showing Biological Activity from the 1975 Literature[a]

Class of Compounds	Total No.	New[b]	Old[c]	Class of Compounds	Total No.	New	Old	Class of Compounds	Total	New	Old
Alkaloids	73	24	49	Monoterpenes	13	1	12	Thiophenes	2	0	2
Sesquiterpenes	47	19	28	Simaroubolides	9	8	1	Sulfides	2	0	2
Diterpenes	26	12	14	Phenolic Acids	8	2	6	Nitro Derivatives	2	0	2
Triterpene Saponins	22	2	20	Amino Acids	8	0	8	Phenylpropanoids	2	0	2
Triterpenes	18	5	13	Lignans	6	2	4	Steroid Saponins	2	0	2
Flavonoids	18	2	16	Carbocyclic Comps.	5	0	5	Cardenolides	2	0	2
Coumarins	15	1	14	Benzenoid Deriv.	4	2	2	Cyano Derivatives	1	1	0
Quinones	15	4	11	Fatty Acid & Esters	4	1	3	Naphthalenes	1	0	1
Sterols	17	7	10	Isothiocyanates	2	0	2	Xanthones	1	0	1

[a]Total number is 325.
[b]Total number is 93.
[c]Total number is 232.

TABLE 7

Numbers of Biologically Interesting Plant Products
from the 1975 Literature by Category of Activity

Type of Biological Activity	No.	Type of Biological Activity	No.	Type of Biological Activity	No.
Analgesics	8	Cardiovascular (Misc.)	5	Hypocholesterolemics	22
Anorexics	2	Cathartics	1	Hypoglycemics	13
Antiarrhythmics	2	Chemother.-Anthelmintics	5	Hypotensives	8
Antiazotemics	1	Chemother.-Antibacterial	30	Immunosuppressants	1
Anticonvulsants	3	Chemother.-Antifungal	28	Insecticides	2
Antiemetics	1	Chemother.-Antimycoplasmics	2	Insect Feeding Deterrants	1
Antifertility Agents	4	Chemother.-Antiprotozoan	30	Psoriasis Ameliorating Agents	1
Antihepatotoxic Agents	3	Chemother.-Antiviral	8	Spasmolytics	7
Antihistaminics	4	Choleretics	5	Spermicides	3
Antiinflammatory Agents	32	Cholinomimetics	1	Teratogens	6
Anti-sickling Agents	4	CNS Active Agents	14	Toxic Plant Principles	12
Antitussives	3	Coronary Vasodilators	1	Tumor Inhibitors	63
Antiulcer Agents	12	Diuretics	4	Cytotoxic Agents	49
Capillary Antihemorrhagic	2	Fish Poisons	1	Tumor Promoters	25
Cardiotonics	1	Gonadotropics	1	Uterine Stimulants	1
				Vasodilators	1
				Miscellaneous	3

issed for substances isolated from higher plants alone.

It may be of some interest to learn also that syntheses
were published for about 1000 natural products, and that the
structures of 275 natural products were determined by X-ray
analysis during 1975.

A total of 325 compounds of known structure, isolated
from higher plants only, were reported in the 1975 literature
as having one or more types of biological activity in some
system(s) having relevance to their potential use as a drug.
Of the 325, ninety-three were compounds of novel structure reported for the first time, and 232 were preciously known
structures (Table 6). As would be expected, the majority of
these biologically active plant principles were alkaloids
(73/325), followed by sesquiterpenes (47/325), diterpenes
(26/325), triterpene saponins (22/325), triterpene aglycones
(18/325), flavonoids (18/325), coumarins and quinones (15/325
each), sterols (17/325) and monoterpenes (13/325).

The various categories of biological activity for these
325 compounds are summarized in Table 7. As can be seen, various categories of chemotherapeutic agents (antibacterial,
antiprotozoan, antifungal, antiviral) were most frequently
cited, followed by tumor inhibitors and cytotoxic agents,
antiinflammatory agents, tumor promoters and/or co-carcinogens,
hypocholesterolemics, hypoglycemics, anti-ulcer agents, and
toxic plant principles.

It was not possible to determine how many of these activities were published from laboratories of pharmaceutical
firms, but it would be safe to estimate that perhaps less than
10 per cent would be categorized as such.

These data do not include a large number of reports concerning interesting biological activities of extracts from
organisms which remain to be studies for their active princi-
ple(s).

VI. SUMMARY

It is apparent from the data presented herein that plants
have served as major and important source of drugs necessary
for the health-care system in the United States, and throughout the world. However, very little effort is being expended
to discover new drugs from natural sources. The reasons for
this apathy have been discussed in detail elsewhere (7). From
data presented, based on the 1975 literature, it is evident
that many unique biologically active compounds are being discovered from plants, but they are not being explored to the
fullest extent as potential drugs for use in man.

VII. REFERENCES

1. Anon. Herbal Pharmacology in the People's Republic of

China, National Academy of Sciences, Washington, D. C., 1975.

2. Hsu, H.-Y. Personal communication, 1977.

3. Chikov, P. S., *Khim.-Farm.Zh.* 10, 97 (1976).

4. Anon., Annual survey report, 1973-1974. Pharmaceutical Manufacturers Association, Washington, D.C., 1975.

5. Anon., National Prescription Audit Ten Year Summary, 1959-1968, and subsequent supplements, R. A. Gosselin Co., Dedham, Massachusetts, and Lea Inc., Ambler, Pennsylvania, 1969.

6. Anon., National Prescription Audit General Information Report, 12th ed., Lea, Inc., Ambler, Pennsylvania, 1972.

7. Farnsworth, N. R. and Bingel, A. S. in "Medicinal Plant Research", (H. Wagner and P. Wolff Eds.), Springer-Verlag, Berlin, 1977.

POTENTIALS FOR DEVELOPMENT OF WILD PLANTS
AS ROW CROPS FOR USE BY MAN

Arnold and Connie Krochmal

Southeastern Forest Experiment Station
U.S. Department of Agriculture
Asheville, North Carolina

The potentials for transferring wild plants from the forest understory to cultivated situations are a barely touched natural and renewable resource.

From 1966 to 1971 while working in Kentucky with the U.S. Forest Service we carried on a program to encourage the domestication of certain useful wild plants. The principal resource of interest at the time was the potential for growing wild plants used for medicinal purposes as backyard row crops. In addition, of course, there are wild plants that serve as food sources and others that are useful as the sources of natural dyes (1-6).

Domestication of wild plants is rarely successful. The last success prior to ours in Kentucky was the domestication of the cranberry.

During our 5-year project three plants were studied extensively: *Lobelia inflata*, Indian tobacco; *Phytolacca americana*, poke; and *Podophyllum peltatum*, the mayapple.

Indian tobacco is the source of an alkaloid, lobeline, that is incorporated into anti-tobacco preparations sold on the market (7-9). A major drug company has asked for assistance in producing commercial crops of this plant as a lobeline source. We began studies with germination and learned that germination was best under short-red light (4). We then worked through nursery culture and handling and conducted field studies of spacing and actual lobeline yields under varying conditions (10-12). We compared the lobeline yield to sugar yield, in that the plants were serving as the carrier for the desired phytochemical (10,11).

Phytolacca americana was considered a potentially valuable crop (13,14). We found information suggesting that the plant's fruit, and perhaps other parts as well, have the

75

ability to inhibit cell division, a matter of interest to medical scientists working on tumor and cancer control. In addition, the young leaves in the early spring age much eaten in
the Appalachian highlands, and are canned in northern Kentucky
as well as southern Ohio. Growers have asked us for help in
overcoming a seed germination problem. Weavers and craftsmen
in the Appalachian area make good use of the purple berries as
the source of a dye.

Our work showed that the principal problem with the seeds
was their impermeable coats, and the scarification increased
germination markedly. We were able to grow this plant successfully as a row crop.

A third plant of interest was the mayapple because it too
had the ability to inhibit cell division (16). Although they
are highly toxic when green, the ripe fruits have been made
into jam. Regretfully we were never able to achieve germination, although we were able to determine that the seeds were
viable. We worked with embryos alone, all to no avail. We
did discover that the ground and powdered fruits at certain
concentrations inhibited germination of other seeds. Using
TLC we extracted five substances, two of which were germination inhibitors when used in water solution.

An interdisciplinary approach was used throughout this
work, with Dr. Leon Wilkens and Dr. Millie Chien of the
School of Pharmacy at Auburn University and Dr. Phil LeQuesne,
then of the Chemistry Department, University of Michigan,
carrying on the phytochemical analyses, and our group working
on agricultural and cultural requirements.

The purpose of these studies was to provide the small
Appalachian landowner with an opportunity to earn a modest
amoutn of supplemental cash income. The crops required only
small plots of land, modest skills, and very small capital
expenditures.

I.   REFERENCES

1.   Krochmal, A., and Krochmal, C., "The complete illustrated
     book of dyes from natural sources:, Doubleday, New York,
     1974.
2.   Krochmal, A., and Krochmal, C., "A guide to the medicinal
     plants of the United States", Quadrangle/The New York
     Times, New York, 1973.
3.   Krochmal, C., "A guide to natural cosmetics:, Quadrangle/
     The New York Times, New York, 1973.
4.   Krochmal, A., Econ. Bot. 22, 332 (1968).
5.   Krochmal, A., "Wildland shrubs: their biology and utili-
     zation", USDA For. Serv. Gen. Tech. Rep. INT-1, 1972.
6.   Krochmal, A., Walters, R. S., and Doughty, R. M., "A guide

to medicinal plants of Appalachia", USDA For. Serv. Northeast. For. Exp. Stn. Res. Pap. NE-138, 1969.
7. Krochmal, A., *Lloydia*, 35,303 (1972).
8. Krochmal, A., Wilken, L., and Chien, M., Plant and
    lobeline harvest of *Lobelia inflata* L. *Econ. Bot.* 26, 216
    (1972).
9. Krochmal, A., Wilken, L., and Chien, M., "Lobeline con-
    tent of *Lobelia inflata*: structural, environmental and
    developmental effects", USDA For. Serv. Northeast. For.
    Exp. Stn. Res. Pap. NE-178, 1970.
10. Krochmal, A., and Huguely, J., *Castanea*, 36, 257 (1971).
11. Krochmal, A., and Magee, K., *Castanea*, 36, 71 (1971).
12. Krochmal, A., and Wilken, L., "The culture of Indian
    tobacco (*Lobelia inflata* L.)", USDA For. Serv. Northeast.
    For. Exp. Stn. Res. Pap. NE-181, 1970.
13. Krochmal, A., "Germinating pokeberry seed (*Phytolacca
    americana* L.)", USDA For. Serv. Northeast. For. Exp. Stn.
    Res. Note NE-114, 1970.
14. Krochmal, A., and LeQuesne, P. W., "Pokeweed (*P.
    americana*): possible source of a molluscicide", USDA For.
    Serv. Northeast. For. Exp. Stn. Res. Pap. NE-177, 1970.
15. Krochmal, A., "Mayapple" *Podophyllum peltatum* L., USDA
    For. Serv. Res. Pap. NE-296, 1972.

RECENT EVIDENCE IN SUPPORT OF THE
TROPICAL ORIGIN OF NEW WORLD CROPS

C. Earle Smith, Jr.

Department of Anthropology and Biology
University of Alabama, University

I. INTRODUCTION

Human interest in the origins of cultivated plants is
persistent and universal. One of the first major scientific
treatments concerning origins of cultivated plants was
DeCandolle's, "The Origin of Cultivated Plants" (1). As one
of the leading taxonomists of his time, he made an outstanding
contribution. Many of the concludions which he drew have been
modified very little since. However, the few bits of archaeological evidence available to DeCandolle included only plant
fragments from Egyptian tombs, material from the Swiss Lake
Dweller sites and some plant remains from butials in the
Peruvian desert. Most of the plant material was without specific data about its source in time and culture and none of it
was dated in the way in which we are now able to date archaeological remains. With the archaeological information now
available to me, it is possible to provide somewhat more specific information on origins for a part of the American crop
plants of native origin (See references and Carbonized Plant
Remains from San Jose Mogote (OS-62); The Plant Remains from
Barrio Rosario Huitzo (B-46); Plant Remains from Guitarrero
Cave, Peru (Pan 14-102); Pre-Ceramic Plant Remains from the
Oaxaca Valley; The Tierras Largas (b-74) Vegetal Remains; The
Vegetal Temains from Abasolo. Unpublished Manuscripts.).
I shall not detail other participants in the development
of information on crop origins, because they have been summarized in several papers, some recently published. We have now
reached the stage of knowledge in which we can confidently
point out the continent of origin of almost all cultivated
plants. However, the region of origin for many cultivated
plants is not known and the ancestral species for many crops
is completely unknown yet.

79

The main focus of the symposium is on chemurgic or industrial applications, if I can judge from the titles. I might
point out that often the potential of old crops has not been
completely realized. Some like the chili peppers would not
seem to have much potential. Other groups of plants, like the
Cactaceae, may be untapped reservoirs of chemical products.

While there are many ways to approach the problem of an
order in which to discuss the American crop plants, none are
completely satisfactory. Because uses overlap, and the order
of importance is difficult to establish once one has discussed
maize, beans and squash, I shall resort to a botanical strata-
gem and discuss American crop plants in the Engler and Prantl
order of classification (2).

II.   FAMILIES OF PLANTS WITH CROPS OF NEW WORLD ORIGIN

A.   Gramineae

This then, brings us to the Gramineae of which the more
important American crop is maize or Indian corn (*Zea mays*).
Since the last time I discussed this crop at an Economic
Botany symposium, no further progress has been made in resolv-
ing, finally, the ancestral species from which maize originated. The place is no longer in much doubt following the recovery of early maize remains in the Tehuacan Valley (3,4). The
crop must have originated in southern Mexico. More recent
archaeological recoveries of remains of maize have been made
in Peru which may date back to as early as 4,000 B.C., but
they hold no real surprises. The earliest cobs from
Guitarrero Complex III are not much different from later
Peruvian maize, but they do remind me of Bat Cave maize. They
are not primitive enough to suggest that origin was any place
except southern Mexico.

Carbonized maize remains have now been studies from the
Chiriqui Province of Panama which place maize here about AD
300. W. C. Galiant (1974, personal communication) feels that
these ears show characteristics of both Nal Tel from Mexico
and Pollo from Colombia. It would seem that this area was
first bypassed when maize was introduced into South America,
but later received maize from both North and South America.

No other American grass has made it into modernity as a
crop, but some evidence is available for the cultivation of
another American grass (28). At the Ocampo Cave sites of
Tamaulipas, Mexico, excavated under the direction of R. S.
MacNeish and from the Tehuacan Valley Caves, also excavated
under the direction of MacNeish, fecal samples were recovered
which were analyzed by E. O. Callen. In the Ocampo Caves,
Setaria geniculata seeds were a principle component of human

meals from about 4,000 B.C. to at least 2,200 B.C. when maize
came into the Tamaulipas area. Foxtail millet was also a
major component of human fecal specimens from the Tehuacan
Caves where the species harvested may have been *Setaria
macrostachya*. In the Tehuacan Valley, *Setaria* was found in
the lower levels of the Coxcatlan Cave deposit and the amounts
increased upward until maize became an effective crop plant.
Thereafter, *Setaria* persisted until Spanish contact. The samples found in the Ocampo Caves show an increase in size upward
in the deposit indicating that the *Setaria* was under cultivation (5,6). Eventually, amize and the European introductions
of food plants forced *Setaria* into disuse and it has not been
observed in cultivation recently.

While it now appears unlikely that we will have a need
for an additional cultivated millet, it is possible that the
Mexican millets can be recultivated and that they might fill
an agricultural niche where currently utilized small grains
cannot be grown.

B.   Palmaceae

Archaeological remains have been recovered for a number
of species of palms. From the Tehuacan Valley sites, many
fragments of coyol (*Acrocomia mexicana*) were found, some of
which date to about 4,800 B.C. when they were introduced into
irrigation cultivation (7). Coyol will not survive the long
dry season of the Tehuacan Valley without supplementary water
and this provides us with a date for the use of irrigation in
the Valley. Coyol is still grown and casually used and the
amount of material recovered from the archaeological deposits
indicate that it has never been heavily used.

In the Chiriqui area of Panama, the fruit of *Acrocomia*
(probably corozo pacora, *A. vinifera*) form a major portion of
the carbonized plant remains recovered from several sites.
The record covers 6,000 + years including the later prehistoric period when the subsistence pattern included maize and
beans. From measurements of the fragments it was possible to
approximate the diameter of the intact fruits. The data
derived from the measurements indicate a range of variability
which suggests that cultivation of palms has been underway for
much of the period. At the present time, palm cultivation is
often so casual that it appears to be unpremeditated. However,
natural populations with freely interbreeding individuals tend
to have fruit size variation within limits which preclude very
large and very small fruit. When artificial selection influ-
ences a crop, human intervention usually preserves the unusu-
ally large fruit sizes by preference.

In spite of the probably long history of cultivation of
the humid tropical forest palms, I know of no concentrated

study of the chemical compositions of the oils or other products from the plants. It is now obvious that many of these
are well beyond the primary stages of selection in cultivation
and that they will probably grow well under plantation conditions. Several species of palms in use in the seasonally dry
areas of Mexico suggest that these might also be worth investigation.

C.   Bromeliaceae

In these days of synthetic fibers, it would seem point-
less to emphasize the use of fibers from bromeliaceaous plants
in prehistoric times. To the best of my knowledge, none of
these plants were ever brought into cultivation. However,
several species of *Tillandsia* and *Hechtia* were obviously used
for the extraction of fiber in the Tehuacan Valley as were
Hechtia, *Tillandsia* and *Bromelia* in the Tamaulipas area (8);
Tillandsia and *Hechtia* in the Oaxaca Valley and *Tillandsia* and
Puya in the Callejon de Huaylas, Peru (Smith, n.d.). With the
future uncertainty of supplies of petrochemicals, we might
once again be reduced to natural fibers and bromeliaceous fibers might be a useful addition to the world's fiber inventory.
Incidently, the us of pineapple by humans is recorded
in the coprolites of upper levels of Coxcatlan Cave in the
Tehuacan Valley (9). No macroscopic remains of *Ananas* were
recognized in the Cave debris. It seems logical to suppose
that these plants were in cultivation in the area inasmuch as
the area of natural geographical distribution would seem to be
the Caribbean costal area of South America.

D.   Liliaceae

While several liliaceous plants have been identified from
sites, none have been claimed to have been cultivated.
Beaucarnea gracilis of the Tehuacan Valley was a source of
leaves used for weaving. *Allium* sp. bulb coats in the Oaxaca
Valley suggest a use of wild onion for flavoring.

E.   Amaryllidaceae

The species of *Agave* have long been exploited for both
fiber and food. Quids, an indication of use as food, are
found in the earliest level of Guila Naquitz Cave (about 8,500
B.C.) and maguey is an important item in human feces from
early to late levels of caves in Tamaulipas and Tehuacan. The
fragments of worked leaves indicate that the extraction of
fiber was long known in these same areas. Unfortunately, the
magueys are vegetatively propagated and the material from
early to late provides no morphological clue to the time of
first cultivation. We can be sure that it was in the Mexican

uplands that this took place, because this is the area of
natural distribution of the wild species from which the cultivates are derived. Many maguey species have been examined
chemically, but more should be investigated inasmuch as many
species of *Agave* succeed well in areas marginal in soil fertility and rainfall.

The South American equivalent, *Furcraea*, is represented
in Guitarrero Cave from the earliest levels to the surface.
It is not edible, according to my informants there and elsewhere, and the leaf material must have been gathered for the
extraction of fiber. Thus, we have an indication of the beginnings for a fiber industry at 8,500 B.C. in Peru, which
later seems to have been a major industry at this site.

F.   Iridaceae

In Peru at the Guitarrero Cave site, *Cypella peruviana*,
mais de perdices, was apparently harvested for the bulbs, but
I have no indication of its use.

G.   Fagaceae

The genus *Quercus* grows in many habitats in the temperate
zone and many of us forget that it is also prominent in some
areas of the tropics. In the eastern deciduous forest area
of North America and in the Oaxaca Valley site, Guila Naquitz
Cave, acorns were a prominent item in the carbohydrate intake
of the populations. In a sense, we have semi-cultivated oaks
for centuries in using them for ornamental and shade trees.
Perhaps we should now consider them as a prominent candidate
for a crop plant for marginal land which will help fill future
food needs.

H.   Lauraceae

The available information on the place of development of
avocado as a cultivated crop still remains centered in the
Tehuacan Valley. Here was found the earliest known avocado
cotyledon (probably about 7,500 B.C.) and it is here that the
series of cotyledons from upper and later levels shows a grad-
ual increase in size. During Classic and Postclassic levels
(AD 300 to 1500), avocado cotyledons as large as those removed
from California avocados today indicate that the Mexican race
of avocado had been stabilized by selection from *Persea americana* var. *drymifolia* (10,11). Avocado pits are known as carbonized remains recovered from Formative village sites in the
Oaxaca Valley in sufficient numbers to suggest that they were
a regular part of the cropping system and the human diet.
They have also been found in late levels in coastal Peruvian
sites where they would have to have been under irrigation.

I.   Leguminosae

A number of leguminous plants have been brought into cultivation in various places in the Americas (29). Some of
these, like the tree legumes which are regularly favored or
planted (but never in orchards), provide no morphological evidence with which to differentiate cultivated from wild trees.
A number of these are locally popular and have not been widely
distributed. Others are used over a wide area, but different
species are used in different areas. An example of the former
is guaje, *Leucaena esculenta*, which is planted and harvested
in southern Mexico primarily by the Indian population. Pod
fragments of this tree were recovered in early levels in the
Tehuacan Valley, they date to 8,500 B.C. in the Oaxaca Valley,
but they have not been recovered from sites outside of southern Mexico. From the pod remains, it is not possible to distinguish gathered pods from pods from cultivated trees.

An example of the widely distributed usage involving different species of the same genus is mesquite. *Prosopis
juliflora* is still widely utilized in Mexico where it may have
been one of the early cultivates. It is in the earliest
Tehuacan and Tamaulipas levels and in the earliest Oaxaca
levels. Chewed quids of mesquite pods and caches of seeds illustrate its use. In Peru, algaroba, *Prosopis chilensis*, is
known from coastal sites and it is similarly used. Again, in
Peru, it is planted in hedge rows or adjacent to buildings,
but we have no evidence from the archaeological remains which
allows a distinction to be made between gathered and cultivated algaroba or mesquite.

Several herbaceous members of the Leguminosae are known
from archaeological contexts. Jack beans, *Canavalia ensiformis*
and *C. plagiosperma*, appear in both Mexican and Peruvian sites.
In Mexico, a single seed is reported from the Tehuacan Valley
at about 3,000 B.C., but a number of *C. ensiformis* seeds were
found in levels dating to about AD 900 in the Oaxaca Valley.
In Peru, *C. plagiosperma* was discovered in Huaca Prieta in
contexts which date it at about 2,500 B.C. Later finds include both species in other sites of the Peruvian coastal
area. Inasmuch as *C. ensiformis* may have been selected from
C. brasiliensis (12), cultivation may have begun in either
North or South America.

The genus *Phaseolus* is of much more economic importance
at the present time and is therefore of continued interest.
Several species have been brought into cultivation of which
two, *P. vulgaris* and *P. lunatus*, are of particular interest.
Two others, *P. coccineus* and *P. acutifolis* var. *latifolius*,
are of lesser interest. A fifth species, as yet unidentified,
provides a note of mystery in the story of bean use.

The unidentified species of *Phaseolus* was discovered in

early to late levels of Guila Naquitz Cave in Oaxaca. Lawrence
Kaplan (Oaxaca Valley Species of *Phaseolus*, unpublished manuscript), who has studied these seeds, found that the embryo
anatomy is that of a species with hypogeal germination, i.e.,
that cotyledons are not raised above soil level on germination.
He feels that the beans were largely gathered from wild plants
rather than being cultivated. While they were an integral
part of the diet for cave inhabitants for a period from about
8,000 B.C. to AD 900, they do not appear to have given rise to
any cultivated bean nor does their use seem to have persisted
until the present time. Thus, we can only surmise that this
represents a very local usage.

On the other hand, *Phaseolus vulgaris*, common bean, is
currently a widely distributed part of the native vegetation
of the Pacific slope of North America from Mexico to Costa
Rica and of similar habitats from Venezuela to Argentina on
the eastern Andean slopes from 1500 to 2800 m elevation (13).
The archaeological record of common bean includes recoveries
from the Tehuacan Valley at about 4,000 B.C., and from the
Tamaulipas area at about the same time period, always as cultivated varieties. In Peru, common bean was recovered from
Guitarrero Cave at levels dating to about 7,000 B.C. securely
and to 8,500 B.C. with some reservation. These were completely cultivated varieties of common beans. Inasmuch as these
were accompanied by specimens of *Phaseolus lunatus*, lima beans,
which were also clearly cultivated varieties, we now must look
to South America as a probable area for the domistication of
common bean and lima bean during a period prior to 8,000 B.C.
I have little doubt that the common bean also came into cultivation in North America, because the wild *Phaseolus vulgaris*
stocks are generally genetically compatible with the old cultivated varieties of Mesoamerica. In fact, the ground color
of the seed coat in regional varieties of common benas is
often very similar to the seed coat color of wild beans collected in the same area suggesting that gene exchange has resulted in a regional color patterning (14). In Mesoamerica,
the cultivated varieties of *P. lunatus* are the small-seeded
forms known as seiva beans which almost certainly originated
from local wild stocks of *P. lunatus*.

It now becomes apparent that the multiple origin of some
cultivated plants is indeed a fact. Both lima bean and common
bean have a very wide natural distribution. The seed sizes of
local varieties of lima bean in archaeological samples con-
firms the multiple origin of this cultigen in Andean South
America and in Mesoamerica. I do not know that chemical similarities have been demonstrated for wild and cultivated common
bean varieites in Mesoamerica, but Brücher (13) reports that
serological and chemical similarities have been demonstrated
for wild and cultivated common bean varieties on the eastern

Andean slopes. Furthermore, the cultivated bean varieties
from Guitarrero Cave suggest that we must revise our timetable
for the beginnings of plant cultivation in the South American
region. Brücher's evidence for wild bean distribution also
emphasizes the accuracy of the hypothesis that cultivation of
plants was initiated east of the Andean highlands and cultivated plants were introduced into the highlands and eventually
into the Pacific coastal region. With the rapid destruction
of wild bean habitats in both North and South America, it
would seem to be a wise move to secure as much germ plasm of
the wild varieties as quickly as possible in order to preserve
them for future use as well as to serve as a base for current
breeding and chemo-resource programs.

The tepary bean, *P. acutifolius* var. *latifolius*, appears
to have a native distribution from the southwestern United
States southward along the Pacific slope of Mexico to about
Oaxaca (15). Prehistoric use of the tepary bean apparently
includes the U.S. southwestern cultures and the inhabitants
of the Tehuacan Valley where it was introduced about 4,000
B.C. Where it might first have been cultivated has not yet
been discovered, but it could not have been the Tehuacan
Valley where it is not native.

The scarlet runner bean is a highland bean which is not
well known in the United States although it is cultivated in
some parts of Mexico. The archaeological record includes recoveries in the Tamaulipas areas dating to about 7,000 B.C.
which Kaplan, and MacNeish (16) have suggested are gathered
remains from plants which were perhaps harvested for an under-
ground storage root rather than for seeds. However, the area
is not known to be a part of the native range of wild
P. coccineus. Other recoveries of this bean were made in the
Tehuacan Valley at about 200 B.C. and in the Oaxaca Valley at
about AD 700. Scanty remains suggest little popularity, but
the seeds are definitely from cultivated plants. In both
valleys, they would be unlikely cultigens and may have been
imported from higher on the surrounding mountains. It is
possible that these beans are evidence of a widespread marketing system similar to that in operation in rural Mexico in
recent times.

In view of the obvious early interest in cultivation of
plants in South America, it might not be wise to dismiss off-
hand the early cultivation of plants in North America. The
out-of-place scarlet runner been in the Tamaulipas area could
be a bit of evidence that this species was an early cultivate
introduced into the Tamaulipas area in cultivation. Later, I
shall examine other evidence which seems to suggest that this
is not impossible.

In view of the knowledge that small, wild bean seeds have
proportionally more protein than the large cultivated varie-

ties of *Phaseolus vulgaris* (Kaplan in reference 17) it may be profitable to examine the seeds of the approximately 100 species of *Phaseolus* for amino acid content. While many of these may be of limited use for direct consumption, amino acids produced in quantity by some may find an important place in synthetic foods for a protein-hungry world population.

J.   Euphorbiaceae

As yet, no one has recovered carbonized portions of *Manihot esculenta*. While some dessicated tubers have been recovered in late contexts from the Peruvian coastal area, early remains of this plant still elude the archaeologists. In examining humas feces from the Tehuacan Valley, Callen (18) identified *Manihot* starch grains. However, we must remember that Mexico is a center of distribution for species of the genus *Manihot* and it is possible that the starch grains of an unexamined species of the genus may have starch grain morphology similar to that of *M. esculenta*. Tuber material recovered from early levels of Guitarrero Cave probably has no material of this species because of the elevation of the Callejon de Huaylas and the lack of native species of the genus in the uplands.

K.   Oxalidaceae

Among the tubers recovered from the Callejon de Huaylas in Peru are a series of tubers which appear to be oca (*Oxalis tuberosa*). Unfortunately, examination of material under the microscope failed to disclose starch grains in most of the samples and the gross morphology is suggestive, but not definitive. Also, we have not yet discovered a clue which will allow us to separate tubers gathered from wild plants and those grown on cultivated plants. The amount of tuber material suggests that the tradition for tuber use in the Andes predates 8,500 B.C.

L.   Malvaceae

The recovery of Gossypium remains from both North America and South America has not changed the hypothesis that *G. hirsutum* originated in the north and that *G. barbadense* originated in the south. The cultivated forms recovered from the Tehuacan Valley at about 3,000 B.C. are still the oldest examples of upland cotton known. The fortuitous discovery of a cotton peduncle among the remains recovered from the Oaxaca Valley conclusively proved that the Mexican material is *G. hirsutum* (19), and the accidental imprisonment of a cotton boll weevil (*Anthonomus grandis*) in a boll lock indicates that cotton was the host for this insect at least as far back as

AD 700 (20).

To the south, cotton recovered from the Ancon-Chillon region of central coastal Peru and datable to about 2,500 B.C.
has been shown through a constellation of characteristics, to
be *G. barbadense* (21). This series shows a developmental
sequence from wild forms of *G. barbadense* toward the cultivated door-yard cottons of western South America. Thus, it can
be concluded that selection in a cultivated crop cotton was
being carried on in Peru. For *G hirsutum*, this evidence has
not yet been discovered. However, the recent Peruvian information should help discount the early hypothesis that the
origin of these tetraploid cultivated cottons involved human
introduction of an Old World cotton into the Americas.

M. Cactaceae

For all of its importance to peoples in the dry areas of
North and South America, species of the Cactaceae have remained unimportant to those with Eurasian traditions. It would
not be surprising to discover that the nopal or tune (*Opuntia*
spp.) were the first cultivated American plant. The important
dry season place of nopal (stems) and tuna (fruit) for human
nutrition is shown in every dry cave or shelter fill in the
habitat of the genus. Abandoned pieces of nopal root and be-
gen to grow without being planted and this could not have escaped the observation of the early gatherers.

The most widely distributed cacti are species of *Opuntia*.
Some species are too heavily laden with bad flavor or they are
mechanically unsatisfactory for food use. Many other species
have completely edible stems and fruits. Some species bloom
sporadically throughout the year so that at least a few fruits
are generally available, while other species have a more
specific season of bloom. In Tamaulipas, in Tehuacan and in
Oaxaca, nopal and tuna have been regular dietary components as
is shown by the macroscopic remains and in the human coprolites. Because the plants are vegetatively propagated, we
cannot determine when cultivation began.

In the Tehuacan area and in Oaxaca, the fruit and stems
of several species of columnar cacti are utilized for food.
Perhaps the most widely used are species of *Lemaireocereus*.
At least three species in the Tehuacan Valley and two species
in the Oaxaca Valley are harvested. Garambullo (*Myrtillocactus
geometrizans*) is both a dry season and wet season resource in
Tehuacan and the related *M. schenkii* in the Oaxaca Valley is
present in upper levels of Guila Naquitz Cave. *Escontria
chiotilla* (jiotilla) is also popular in the Tehuacan Valley.
These plants appear to provide a regular source of vitamins
and minerals even in the dry season when little foliage is
available for greens.

At the present time, many of the cereus-type cacti are
utilized as cultivated plants, frequently being cut and im-
planted side by side to form a living fence. The fruit which
develops on these is harvested when it is available. Again,
because of vegetative propagation, we have no marked morphological change which provides a clue to the date of first cultivation of these species.

In Peru, the prehistoric inhabitants of the Callejon de
Huaylas used both *Opuntia* and the local species of *Trichocereus*
which grow on the slopes of the Cordillera Negra. Somewhat to
the north of this area, *T. pachinoi* is utilized for its hallu-
cinogenic drug content.

Altogether, the ease of propagation of many species of
cacti and their ability to survive with marginal rainfall
would seem to make them prime subjects for experimentation as
sources of foodstuffs and chemicals for cultivation in marginal areas of the world. Unfortunately, some of the species of
Opuntia introduced into Old World areas have gained for themselves a bad reputation because they have become weedy. At
the same time, cultural bias has often prevented their effective use by local human populations. The latter problem can
probably be ameliorated during times of acute stress such as
the savannah fringe of the Sahara has been suffering in recent
years. The former problem is a biological problem which may
have to be met with a special program to modify the germin-
ability of *Opuntia* seed so that bird and animal distribution
of cacti has proven to be easy: it is only the exploration
for uses which must be pursued.

N.   Sapotaceae

For the most part, fruit comprises only a minor component
in the plant remains recovered either as dried material from
cave and shelter sites or as carbonized fragments from open
sites. However, they obviously played an important part in
human nutrition as the source of vitamins and minerals not
obtainable through the carbohydrate and protein sources. I
have bypassed several families in which fruits were used in
prehistoric times, because they have limited use outside the
tropical areas of the world and because they seem to have
little potential beyond their fruit use. However, members of
the Sapotaceae, while they are largely confined to the tropics,
seem to have a potential beyond their mere use as fruits. At
least one, cosahuico (*Sideroxylon* cf. *tempisque*) appeared in
the Tehuacan Valley at an early level in the caves. Measure-
ment of the single seeds indicated that the fruit increased in
size upward in the deposit showing that artificial selection
pressure was influencing seed and fruit size and that the
trees were under cultivation (22). The latter was already

suspected, because the species is not known to survive the dry
season in the Tehuacan Valley without supplementary water.   In
Peru, lucuma (*Pouteria* cf. *lucuma*) was an important fruit to
the inhabitants of Guitarrero Cave.   Unfortunately, the seeds
were all broken before recovery, apparently by human use of
the endosperm, so that they could not be measured.   In both
cases, the wild trees may have been members of the gallery
forest along the permanent rivers draining the valleys before
human disturbance moved them into cultivation.

O.   Ebenaceae

The persimmon of temperate North America (*Diospyros
virginiana*) was widely used for the making of bread and beer
according to contact accounts (23).   The black sapote
(*Diospyros digyna*) was apparently one of the most popular
fruits of its area in prehistoric time.   Callen, (9) reports
that it was one of the principle ingredients in human fecal
samples from the Tehuacan Valley.   It must have supplementary
water during the dry season and the appearance of the fruit in
cave deposits at about 4,000 B.C. indicates that it was introduced in cultivation and that the use of irrigation was then
understood.

The fruits of these two families seem to me to be potentially important because they have flesh which may be preserved for later use by drying.   Temperate persimmons were often
stored in this way.   Furthermore, black sapote has a high
content of calcium, phosphorus, vitamin A and vitamin C (24)
all of which tend to be needed by people living in poverty
areas of the world.   Species of *Sideroxylon* and *Pouteria*,
other than the species discussed here, have been shown to have
an even higher concentration of the above minerals and vitamins.   All of the species are currently under casual cultivation and they should be selected especially for varieties high
in those nutritional elements which they concentrate.   They
are a potential crop for tropical areas in which other crops
are not now produced for world trade and they might become
important in the impending world food crisis.

P.   Solanaceae

Several members of the Solanaceae are important world
crops at the present time.   Most of these originated in the
New World.   Perhaps the most widely used are members of the
genus *Capsicum* of which four or five species are in regular
cultivation and of which *C. annuum* is the most widely distributed and used.   *Capsicum annuum* (chili) is present in the El
Riego Horizon in level XIX which seems to be datable to
around 6,000 B.C.   From the Callejon de Huaylas, a whole
pepper was recovered from the earliest plant-bearing level

datable to about 8,500 B.C. This specimen appears to be
C. *chinensis* and the attached calyx indicates that it was cultivated (non deciduous). Pickersgill (25,27) reports the
presence of C. *baccatum* in the lower levels of Huaca Prieta on
the Peruvian coast. She has localized the area of origin of
this species in southern Peru and Bolivia and it must have
come to Huaca Prieta as a cultivate.

The regions of origin of the *Capsicum* species have now
been suggested by Pickersgill (25,27). The wild or feral chili
is widely distributed from Florida well into Mexico around the
Gulf of Mexico. It seems apparent that C. *annuum* was first
cultivated in North America. The earliest widely distributed
South American cultivate is C. *baccatum* var. *pendulum* which
probably originated in the area of souther Peru and Bolivia.
Perhaps the earliest pepper in cultivation is C. *chinense*
which probably is derived from C. *frutescens* growing naturally
in the upper Amazon basin east of the Andes and this one seems
to have been accompanied by ancestral C. *frutescens* at about
2,000 B.C. at Huaca Prieta. Thus, the general outlines of the
areas of cultivation of the chili peppers are now known.

Q.   Cucurbitaceae

The genus *Lagenaria* is among the oldest cultivated plants
of the New World. However, the bulk of the related species
are found in South Africa and the ancestral form of
L. *siceraria* is found in South Africa. The forerunner of the
bottle gourd must once have been widely distrivuted and it may
have come into cultivation in the Old World and the New World
concurrently and independently (26).

On the other hand, *Cucurbita* is found only in the New
World as wild species. Interestingly enough, the five cultivated species seem to have had their base in southern Mexico
and Central America, but two of the species may have come into
cultivation in South America. So far as the winter squash,
C. *maxima*, is concerned, we have little new evidence. It
seems to have originated in southern Peru-Bolivia and spread
northward in the Andes and into the coastal valleys. Similarly, C. *ficifolia* may have originated in the Andes and spread
northward to Mexico. It is present in the upper level of
Guila Naquitz Cave, Oaxaca, at about AD 700, but the oldest
record for the cultivate is about 2,500 B.C. at Huaca Prieta.

The oldest of the cultivated squashes is now apparently
C. *pepo*. A seed of C. *pepo* was recovered from level D of
Guila Naquitz Cave and it can be dated at about 7,800 B.C.
from an associated radiocarbon date in the same unit (K.
Flannery, personal communication). Inasmuch as the nearest
related wild squash is C. *texana* from the area of Texas, C.
pepo could only have been introduced into the Oaxaca Valley as

a cultivated plant. This recovery suggests that the Ocampo
Cave seed of Tamaulipas may not be that of just a weedy plant,
but from fruit harvested from plants under primitive cultivation. This also suggests that we may have to reorient our
assumptions concerning the areas of earliest experiment in
cultivation to include parts of the world considered less
sophisticated and less likely to initiate innovations. The
natural area of distribution of *C. texana* and the very similar
C. pepo var. *ovifera* may have been larger 9,000 years ago, but
it is unlikely that it would have estended far enough to have
included any of the southern Mexican region which was later to
flouresce. We might more confidently admit that the inhabitants of the Texas-northern Mexico region of distribution of
C. texana had made the appropriate observations which suggested to them that they might encourage this cucurbit to grow
where it would be more convenient for them to harvest.

III. EARLY AGRICULTURAL TECHNIQUES

An intimacy with modern techniques of cultivation has
largely blinded anthropoligists and others to the extreme
simplicity of the crude, early agricultural techniques. It is
doubtful that more than an observation of the habitat of a
species is needed to ensure successful artificial planting of
most species of plants. The only other prerequisite would be
an understanding of the connection between a seed and a subsequent plant, an observation which must have been made commonly by gatherers. Any seed accidentally moistened and left
without processing may germinate, sometimes within a surprisingly short time. Therefore, that seeds artificially distri-
bured in an appropriate place would increase the amount of
harvestable seeds and made harvesting that much easier should
have been seen by many people. After all, the most elementary
form of cultivation is removing competing vegetation in the
area of a desirable plant (this accomplishes soil aeration as
well as reducing competition for light, minerals and moisture).
The earliest cultivators could not foresee the box canyon
into which they were driving their descendants. This simple
process of favoring more desirable plants obviously turned on
the perpetrators and eventually forced them into full time
food production from cultivated plants. At some point along
they way, the time of the gatherer became absorbed in the enlarging need for tending cultivated plants and the economic
investment then made settlement mandatory. None of this
occurred very quickly as we are now beginning to realize from
the archaeological plant remains.
The biological changes governed by the genetic constitu-
tion of a plant do not happen rapidly except via gene muta-

tions. Only after the establishment of plants in cultivation were humans sufficiently in contact with the same plants over a long enough period of time to catch and protect the favorable mutations. Thus, the mutations which resulted in tough rachises in wheat and barley, non-explosive bean pods, nondeciduous chilis would only have been saved in cultivation, but, at the same time, they increased the bonds which tied early cultivators to their folly. Whenever a monstrous change occurred which resulted in six-rowed barley, multi-rowed ears of maize, greatly enlarged tomatoes, the crop plants had increased value, but they could no longer be successful under natural conditions. Man was irretrievably tied to cultivation as a means of subsistance and he was further tied to a single community during the spring and summer by the demands of the cropping system and in the fall and winter to protect the stored harvest.

The added portion of the equation was the lifting of the restraint on human reproduction afforded by plant cultivation. Lack of mobility due to the demands of the cultivated plants made the production of children at close intervals a burden only on the women. Once the overproduction of children became an established part of community life, the very increased number of hands available for labor in the fields or for doing household chores may have reinforced the overproduction or even stimulated overproduction. This, in turn, made further irrevocable the turn to agrarianism, pastoralism or a combination of these.

We now are at a point in the development of agriculture and the world's population where every available resource will have to be exploited. The archaeological record of cultivated plants in the Americas includes clues to areas where may be found diversified germ plasm. It further emphasizes useful plants which have largely been ignored by those of us having a Eurasian cultural background. Much of the development of cultivation in North and South America occurred in regions of marginal rainfall with plants well adapted to the area. These suggest that exploration of the ancient crops of semi-desert areas might provide useful chemical or nutritional products which can be mass produced in marginal areas all over the world. It is doubtful that early man in the Americas missed much that was available in the way of gatherable food and we probably can do little better on the selection of species adaptable to agriculture. In many instances, the related wild species may provide commercially interesting industrial resources in plants which may be a amoenable to cultivation as their edible relatives. It is imperative that we begin to explore the legacy left us by early humans.

IV. REFERENCES

1. DeCandolle, A.,"Origin of Cultivated Plants", Eng. Trans. New York, 1885.
2. Engler, A. and Diels, L., "Syllabus de pflanzenfamilien", Ausl. 11., Berlin, 1936.
3. Mangelsdorf, P. C., MacNeish, R. S., and Galinat, W. C., Bot. Mus. Leaflets, Harvard Univ., 22, 33 (1967).
4. Mangelsdorf, P. C., MacNeish, R. S., and Galinat, W. C., in "The Prehistory of the Tehuacan Valley", (D. S. Byers, Ed.), Vol. 1, p. 178, University of Texas Press, Austin, 1967.
5. Callen, E. O., Amer. Antiquity 32(4), 535 (1967).
6. Callen, E. W., "Analysis of the Tehuacan coprolites". in "The Prehistory of the Tehuacan Valley", (D. S. Byers, Ed.), Vol. 1, p. 261, University of Texas Press, Austin, 1967.
7. Smith, C. E., Jr., in "The Prehistory of the Tehuacan Valley", (D. S. Byers, Ed.), Vol. 1, p. 220, University of Texas Press, Austin, 1967.
8. MacNeish, R. W., Trans. Amer. Philos. Soc., Vol. 48, 1958.
9. Callen, E. O., "Plants, diet and early agriculture of some cave dwelling pre-Columbian Mexican Indians", Actas y Memorias del Congrso Internacional de Americanistas, Tomo II, p. 641, 1968.
10. Smith, C. E., Jr., Econ. Bot. 20 (1966).
11. Smith, C. E., Jr., Econ. Bot. 23 (1969).
12. Sauer, J., and Kaplan, L., Amer. Antiquity 34(4), 417 (1969).
13. Brücher, H., Ange. Botanik, 42, 119 (1968).
14. Gentry, H. S., Econ. Bot. 23, 55 (1969).
15. Kaplan, L., in "The Prehistory of the Tehuacan Valley", (D.S. Byers, Ed.), Vol. 1, p. 201, University of Texas Press, Austin, 1967.
16. Kaplan, L., and MacNeish, R. S., Bot. Mus. Leaflets, Harvard Univ., 19, 33 (1960).
17. Smith C. E., Jr. (Ed.), "Man and His Foods", University of Alabama Press, University, Alabama, 1973.
18. Callen, E. O., Econ. Bot. 19(4), 335 (1965).
19. Smith, C. E., Jr., and Stephens, S. G., Econ. Bot. 25, 160 (1971).
20. Warner, R. E., and Smith, C. E., Jr., Science 162, 911 (1968).
21. Stephens, S. G., and Moseley, M. E., Science 180, 186 (1973).
22. Smith, C. E., Jr., Econ. Bot. 22, 140 (1968).
23. Swanton, J. R., "The Indians of the Southeastern United States", Smithsonian Institution, Bureau of American Ethnology, Bull. 137, Washington, D.C., 1946.

24. Leung, W. W., and Flores, M., "Food Composition Table for
     use in Latin America". The Inter-departmental Committee
     on Nutrition for National Defense, Nat. Inst. of Health,
     Bethesda, Maryland, 1961.
25. Pickersgill, B., in "The Domestication and Exploitation
     of Plants and Animals", (P. J. Ucko & G. W. Dimbleby,
     Eds.), Aldine, Chicago, 1969.
26. Heiser, C. G., Jr., in "Tropical Forest Ecosystems in
     Africa and South America: A Comparative Review", (B. J.
     Meggers, et al. Eds.), Smithsonian Institute Press.,
     Washington, D.C., 1973.
27. Pickersgill, B., Amer. Antiquity 37, 97 (1972).
28. Callen, E. O., in "Man and His Foods", (C. E. Smith, Jr.,
     Ed.), University of Alabama Press, University, Alabama,
     1973.
29. Kaplan, L., in "Man and His Foods", (C. E. Smith, Jr.,
     Ed.), University of Alabama Press, University, Alabama,
     1973.

REQUIREMENTS FOR A GREEN REVOLUTION

G. F. Sprague

Department of Agronomy
University of Illinois
Urbana, Illinois 61801

The term "Green Revolution" was coined to dramatize the early success achieved with the new rices developed at IRRI in the Philippines and the short-statured wheats from CIMMYT. Fed by this initial success, by concern for the malnourished people of the world, and by a widespread disregard of the biological, economic, political and social complexities involved, expectations for the Green Revolution became highly unrealistic. Nobel laureate Dr. Norman Borlaug, one of the prime movers in this development, gave little support to this extreme optimism. He stated, "The Green Revolution has won a temporary success in man's war against hunger and deprivation; it has given man a breathing space. If fully implemented, the Revolution can provide sufficient food for sustenance during the next three decards. But the frightening power of human reproduction must also be curbed; otherwise the success of the Green Revolution will be ephemeral only." In other words, the Green Revolution represented merely the first installment or down payment on a new system of agriculture for the developing countries. If the full benefits were to be realized these countries must keep up the payments; the payments consisting of an expanded research base and the necessary political, economic and social changes required to permit movement away from a subsistence agriculture.

The desirability of a Green Revolution has been widely recognized, but the requirements for the widespread utilization of the capacity for increased production have been either underplayed or largely ignored. Local research competence is required; one does not just transplant a new and complex agricultural system. Given the required level of research competence, political, economic and social problems must be resolved if a meaningful transition is to follow. Time will not

permit a review of these complex problems, and I shall limit
my remarks primarily to biological requirements which directly
or indirectly affect adoption and utilization of the new potential.

The biological basis of the Green Revolution did not ori-
ginate with the coining of the descriptive term. Rather, it
had its basis in the long series of research developments
within the temperate region; developments involving the fields
of genetics and plant breeding, soil fertility and crop production, plant pathology and entomology. Since temperate
varieties and production practices are generally poorly suited
to tropical areas a direct transfer of technology was unpro-
ductive or unprofitable. The new races of IRRI, the short-
statured wheats of CIMMYT, the new maize hybrids of Kenya, and
the wheat-Medic rotation of the southern Mediterranean region
are the direct outcome of utilizing basic principles developed
in temperate areas for the solution of problems in the new
tropical and semi-tropical environments. In this new situation the basic principles remained the same as did the requirements for adoption and utilization of the newly developed technology.

Perhaps a brief review of U.S. experience would be justified; hybrid corn provides a useful example. In the pre-hy-
brid days U.S. corn yields never exceeded 31 bushels per acre.
The nutrient requirements, particularly nitrogen, were supplied through applications of barnyard manure and the use of
nitrogen-fixing legumes as one component in the crop rotation
system. Mechanical cultivation provided the only means of
weed control. Other than crop rotation and a very limited use
of seed protectants, there were no controls for either disease
or insect pests.

With the adoption of hybrids, yields increased some 25-
30%. The use of hybrids was a single-trait substitution,
other elements of the production system remaining unchanged.
Following World War II nitrogen became readily available, in
forms easy to apply and at reasonable prices. It was rapidly
learned that some genotypes (hybrids) could make effective use
of added nitrogen, other could not. Within the group which
could utilize added nitrogen effectively, it was found that
some genotypes could respond to increased density of planting,
this still further increasing the nitrogen responses. As a
result planting densities have more than doubled and nitrogen
application has increased several fold. With increased yields
per acre it became economically feasible to use an expanding
array of herbicides and to use chemical control for certain
disease and insect pests. Marked improvements in equipment
for soil preparation, planting and harvesting had occurred
throughout this period. Another significant but seldom mentioned development included a marked expansion in a group of

related services which, collectively, can be designated agri-
business. Thus the 1976 production package for corn is markedly different than 1930, a difference related to a nearly
four-fold increase in average yield per acre. Any sustained
Green Revolution suitable for tropical or semi-tropical
regions will require comparable changes in production practices from the traditional to modern or scientific.

Many have assumed that the major component of a Green
Revolution is the development of high yielding varieties (HYV).
The use of the term HYV may be a shorthand convenience but if
taken literally can be quite misleading. In its simplest form
plant breeding involves the development and evaluation of an
array of genotypes and from this array the identification and
propagation of the most productive. The identification of
genetic differences is a complicated process. Productivity is
the resultant of the interaction between the individual genotypes and the environment. Limitations in either genotypes or
environment may establish a ceiling for productive capacity.
If any component of the environment (water, fertility, etc.)
is less than optimum then this limiting factor will determine
yield levels rather than the genetic differences among the
strains under observation or test. A high yielding variety
therefore is one which has the capacity to respond to improved
environmental conditions. If production practices are such
that the environment is limiting, varieties with high production potential may yield but little more than the traditional
varieties in common use. Within rather wide limits the degree
of response is related to the improvement of the cultural environment.

Either land race or improved varieties normally exhibit a
rather limited range of adaptation. This limitation may be
imposed by day-length requirements; insect or disease resistance or tolerance; length of growing season as imposed by
climatic factors, usually rainfall distribtuion or water
supply; and, in some, cases, by the food preferences of the
cultivator. The short-statured spring wheats from CIMMYT are
the major exception to this general rule. These types were
developed with alternating generations grown under different
day length regimes. Therefore the high performing survivors
of the selection process are relatively insensitive to variation in day length. In consequence they have been successfully utilized in Mexico, the southern Mediterranean countries,
the Near East, and the Asian sub-continent. In contrast, IR8,
the first improved rice variety released from IRRI, was much
more restricted in its adaptation. It was too late for
southern China and parts of the rice growing area of Southeast
Asia and India. It performed very poorly in West Africa. In
areas where it was adapted, consumers often faulted it for
quality. The maize hybrids of Kenya are best adapted to the

highland areas of East Africa. These examples illustrate that
even improved types are not universally adapted. Breeding
programs, therefore, must recognize the differing ecological
conditions under which the crop will be grown.

There are many advantages arising from the several international centers; IRRI, CIMMYT, CIAT, IITA, ICRASAT, etc.
They can perform important services in the collection and
regional evaluation of germplasm, the formation of germplasm
pools, in identifying sources of resistance to the important
disease and insect pests to name only a few. As we have just
noted they can also perform useful breeding functions. It is
quite apparent, however, that if maximum genetic progress is
to be achieved, plant breeding and genetic competence must be
promoted at the national level within the developing countries.

As productivity is the resultant of the genotype x environment interaction, equal attention must be given to the
improvement of cultural practices or to the entire farming
system. Fertility and water control are usually the most
limiting factors. Timeliness of planting and plant densities
are a close second in importance. In traditional agricultural
systems maintenance of fertility has received little attention.
Improved varieties, if effective, are merely more efficient
soil depletors. If increased yields are to be achieved and
sustained, the elements required by the growing crop must be
supplied in proportion to the expected yield.

The plant nutrients required to maintain productive capacity at any desired level are dependent on the native characteristics of the soil, cropping or management history, climate
conditions, the crop being grown, and many other factors.
Nitrogen is almost universally limiting. The requirements for
the addition of other major elements, phosphorus and potassium,
may vary with the parent rock from which the soils were derived as well as period under cultivation and the management
practices used. Even if current supplies are adequate heavy
cropping patterns will eventually require, as a minimum,
replacement additions. In acidic or highly leached soils some
minor elements become relatively unavailable, thus drastically
limiting crop production. The required additions may be small
but correction of the problem is essential to a productive
agriculture.

The establishment of nutrient requirements of individual
crops, the response curves of applied nutrient or nutrient
combinations and the interactions with varieties, time of
planting and population densities tend to have a high degree
of area specificity. Research of the type and extent required becomes a national obligation. The international centers can play only a limited role in this aspect of the total
development.

If acceptable, short-term answers have been found in

terms of responsive varieties and improved production and management practices. Then many other requirements for improved
production will be found to be out of phase with the increased
potential. Some of these are ancillary to increasing production; other are basic.

As stated earlier the term "high yielding varieties" is
something of a misnomer. Rather, the new varieties have a
greatly improved capacity to respond to environmental improvement. Without environmental improvement the increase in yield
potential is not sufficiently obvious to be persuasive to the
subsistence cultivator.

A subsistence agriculture has little need for a seed industry. However, if imporved varieties are to successively be
introduced, as technology will inevitably make possible, a
seed industry becomes indispensable. In the developing countries there has been a strong tendency for seed organizations
to be established under government control or sponsorship. In
most instances, to my knowledge, such operations have been
highly inefficient. Several authors have cited comments from
Indian farmers to the effect that seed of introduced wheat
varieties is markedly superior to seed produced locally. I
have seen many instances of extensive varietal mixtures in
corn, sorghum and wheat. Good quality seed, readily available
at sowing time, is a necessary requisite of the new technology.

Improvements in the environment may require the use of
fertilizers, improved water management and any necessary use
of herbicides, fungicides, or pesticides. Again the required
inputs must be available when needed and in the required
amounts. If application of any of these inputs is delayed the
expected benefits will be greatly reduced or vanish. In our
system of agriculture agri-business performs these many services in a competitive and efficient manner.

The Extension Service has made an important contribution
to progress in the United States. It appears that a comparable function needs to be performed within the developing
countries. However, it is easy to overlook the reason for the
effectiveness of extension activities. Basically extension is
an intermediary in the flow of new information from the research centers to the farmers. This information covers the
entire field of agriculture interests: new varieties, improved management practices for both crops and livestock, improved
methods of harvesting and storage to name only a few. The
efficiency of the transfer process is dependent upon many
factors, but one very necessary prerequisite is the continuing
supply of basic information provided by research. Without a
strong research base extension has no relevant information to
transfer.

The elements discussed thus far as requirements for a
continuing Green Revolution are either biological or have an

important biological component. Other factors may have a
strong influence on the extent to which the new potential is
realized. All aspects of the Green Revolution require an increased level of economic inputs; new seed, fertilizer, etc.
Certainly the price paid for the final produce must cover the
cost of inputs plus some additional margin for the extra labor
and management that may be required. Obviously, if prices of
agricultural produce are set at levels which do not cover the
full package of practices adoption will be proportionately
reduced.

If increased production is achieved, deficiencies in the
storage, transportation and marketing system may become apparent. An experience in Nigeria serves to illustrate the
problem. An intensive extension effort was conducted in the
western region. A simple package of practices was promoted
involving several villages. The practices included seed of
viability, proper time of planting and proper plant densities,
the package involving no increased cash outlay. The extension
effort was highly successful, yields of the participating
villages being increased 50 to 100%. Surveys the following
season indicated a complete return to the traditional system.
Inquiries established that as there had been no market for the
increased produce the improved package of practices had no
value. At the same time that the participating villages had
an unsalable surplus, other areas in Nigeria had severe
shortages. Due to a primitive transportation and marketing
system, surplus or scarcity of food staples was entirely a
local problem.

The Green Revolution, in practice, has been criticized
because it favors the innovator, and the innovator is most
likely to be the larger landholder with more capital and in a
better position to survive any possible adverse outcome.
Numerous studies have shown the importance of the innovator in
the adoption of any new development, in agriculture or in
industry. The hope that increases in productivity would solve
social imbalance has no precedent in history. Imbalance in
income distribution is primarily a political and economic
rather than an agricultural problem.

We have considered, very briefly, some of the direct and
indirect biological inputs necessary to initiate and sustain a
"Green Revolution". Obviously many other system modifications
will also be required. We have alluded to the importance of
adequate storage, transportation, and marketing systems.
Credit is usually unavailable to the subsistence farmer and
without credit new innovations become difficult or impossible.
In some cultures, land tenure systems are not conducive to
agricultural progress. Government policies controlling prices
of food grains may simplify the problem of feeding the urban
poor but at the same time limit the total produce available

for allocation. Lack of suitable price differentials may also
contribute to a worsening nutritional balance, e.g. the
increase in cereals and decrease of pulses in India. Literature on these and other potential limitations is extensive.
Our purpose in listing the above few is to emphasize that a
technology which satisfies the biological requirements without
consideration of relevant political, economic, or social
issues cannot achieve its maximum potential.

HOW GREEN CAN A REVOLUTION BE

Jack R. Harlan

Crop Evolution Laboratory
Agronomy Department
University of Illinois
Urbana, Illinois 61801

The story of the green revolution has been told so often, there is no need to repeat it here in any detail. The Mexican model is most instructive. In 1945 at the end of World War II, Mexico was importing 15-20% of her basic food grains. A project was launched jointly by the Rockefeller Foundation and the Mexican Ministry of Agriculture. Work got underway, concentrating on increasing the production of wheat and maize, the major cereal crops of the country. Additional programs dealt with other basic crops. In due time, these efforts bore fruit in spectacular fashion, Table 1.

TABLE 1

Some Production Statistics for Mexico,1950 and 1970 (1)

	1950	1970	Increase Factor
Wheat Production (tons)	300,000	2,600,000	8.7
Wheat Yield (kg/ha)	750	3,200	4.3
Maize Production (tons)	3,500,000	9,000,000	2.6
Maize Yield (kg/ha)	700	1,300	1.9
Bean Production (tons)	530,000	925,000	1.7
Sorghum Production (tons)	200,000	2,700,000	13.5

By the mid-1960's, Mexico was no longer an importing nation but an exporter. During the years 1964-1969, exports were on the order of 5.4 million tons of maize, 1.8 million

tons of wheat and 0.34 million tons of beans. The program
was, by any measure, a tremendous success and gave hope to
developing countries around the world. The high yielding
"Mexican wheats" were introduced to many developing wheat-
growing countries in Asia, Africa and the Americas. Spectacular results were obtained as in Mexico.

In the early 1960's, the International Rice Research Institute was established in the Philippines and staffed with
scientists who set out to do for tropical rice what N. E.
Borlaug and his associates had shown could be done for wheat.
Results were not long in coming and high yielding tropical
rice cultivars were soon available for developing rice-growing
countries around the world. These also were, often, spectacu-
larly successful. It looked as if the world's food problems
were finally going to be solved. Enthusiasm was general,
hopes and expectations high. The movement was called the
"green revolution." Borlaug was, quite justly, awarded a
Nobel Prize for his contributions. After a record harvest,
Prime Minister Ghandi declared that India had achieved self-
sufficiency and would never again need to import food grains.

Two years later, India was importing food grains. In
the early 1970's, the Mexican agricultural boom began to run
out of steam. Today, Mexico is right back where it all began,
importing 15-20% of her food grains. After all that hard
work and all that effort and all those grand expectations,
these results must be classified as disappointing to say the
least. Actually, Mexico is worse off, strategically, than
before it all began. In 1945, the population was 22 million,
today it is about 60 million. There is no way to go back.
Green revolution technology must be pushed as far as possible
simply to reduce imports, let alone feed the people. In 1945
the possibility of a green revolution was still ahead for
Mexico. Now it is already installed and that option is no
longer available. The same is or soon will be true for many
developing countries.

The so-called browning of the green revolution has also
received a lot of public attention. The pendulum swung in
the other direction. Some condemned the green revolution outright as a hoax perpetrated by the rich nations on the poor or
by the rich and middle classes on the destitute, a fraudulent
way to foster western technology and business and to reap pro-
fits from the underdeveloped, a new kind of colonialism and
so on. The new technology, it was said, was energy intensive,
polluted the environment and disrupted social and economic
systems. Genetic resources were eroded and the crops became
genetically vulnerable. A good deal of emotion and not too
much light have been generated on the issues. It is some ten
years since the early enthusiasm and it should now be possible

to put green revolution technology in perspective and appraise
it realistically.

First of all, we should note that the green revolution
is not an exotic phenomenon that happens only in Mexico or
India. It is merely an application of modern technology to
agriculture and is a phenomenon common to all the developed
nations. A green revolution could be defined very precisely
in mathematical terms should one choose to do so. Figure 1

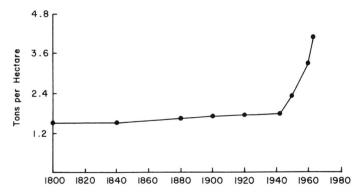

Fig. 1.   Maize yields per hectare in the U.S.

shows a smoothed curve for maize yields per hectare in the
United States. From the beginning of statistical recording
until 1942, increase in yield per hectare was trivial if significant at all. After 1942 the yield curve climbed steeply
and may still be going up. The green revolution for maize in
the U.S. could be described by a ratio of the line before
1942 to the slope of the line after 1942. Other crops show
similar curves, varying only in slopes and break points.

Green revolution curves for Western Europe are very similar in conformation to the U.S. maize curve but the upward
limb of the graph usually began after the end of World War II.
Figure 2 shows yield per hectare for rice in Japan. The slope
is the same, but the sharp increase started near the turn of
the century, well ahead of the rest of the world. A bobble in
the curve is conspicuous and represents the effect of World
War II.

The connection between yields per hectare and the last
world war is very direct. It was during that tragic period
that the developed nations acquired the technology and capacity to fix nitrogen on a large scale. This was for the purpose of manufacturing explosives, but after the war was over,

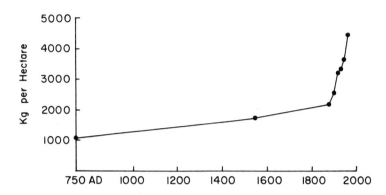

Fig. 2.   Rice yields per hectare in Japan

the technology could be turned to the manufacture of nitrogenous fertilizers. Cheap, fixed nitrogen, and lots of it, is
the foundation of green revolutions.

    To fix nitrogen by any means, industrially or biologi-
cally, requires a lot of energy. Industrially, we have mostly
used natural gas in the United States. Production of this resource is declining. The OPEC nations have served notice that
the time of cheap energy is over. The era of cheap fixed nitrogen ended at the same time. Green revolution technology
is going to be expensive. Countries lacking their own abundant sources of energy will need to make some hard decisions.

    Energy and fixed nitrogen are only two ingredients of a
green revolution. The fertilizer is of no use without appropriate, adapted, fertilizer responsive, high yielding cultivars. In cereals, they are usually semi-dwarf in stature to
resist lodging. Good levels of resistance to the local species and races of diseases is essential.

    The last characteristic is one of the most critical of
the whole system. One specific example can illustrate the
problem. Upper Volta is located in a savanna zone and the
primary food grains are sorghum and pearl millet. The Banfora
plain, however, is traditionally a rice-growing area. African rice (*Oryza glaberrima*) was grown from ancient times but
has been largely replaced in recent decades by Asian rice
(*O. sativa*). The traditional yields are on the order of 700
kg/ha. In the late 1960's a green revolution technology was
installed through international assistance of various kinds.
A dam was constructed on one of the branches of the Volta;
land was leveled and ditched for irrigation. High yielding
cultivars and nitrogenous fertilizers were imported. Yields
jumped to seven tons per hectare, and, with irrigation, two

crops were possible for a total of 14 tons per hectare.   The
farmers were ecstatic.
      Then came 1971.   The new cultivars were hit by blast
(*Piricularia oryzae*).   The yields were right back where they
started at 700 kg/ha.   But, now, nobody could afford that sort
of yield.   After investing in the dam, the irrigation project,
the seed and the fertilizer such yields are no longer toler-
able.   Seeds of new cultivars, hopefully resistant to blast,
were flown in from the Philippines.   This not only was expensive but could only provide temporary relief since the West
African races of blast are probably different from those in
the Philippines.   The only long range solution to problems of
this nature is a strong, indigenous phytopathological and
plant breeding research program.   This is also expensive and
likely to be beyond the resources of Upper Volta to support.
      The new high yielding cultivars are, in fact, genetically
vulnerable if they are developed in one region and used in
another.   The new technology requires not only industrially
fixed nitrogen, but many teams of scientists including plant
breeders, plant pathologists, plant physiologists, nematolo-
gists, entomologists, soil fertility and soil management specialists and so on.   The developed countries have been able to
supply these, but developing countries are likely to have
other priorities for scarce resources.
      National priorities have a great deal to do with success
or failure of green revolution technology.   In the general
rush to industrialize, agriculture is too often neglected and
nations suffer the consequences.   The technology is expensive,
as we have seen, and the farmer must have access to credit at
reasonable rates.   This is often as important as any other
ingredient in the system.
      An orderly marketing system is also essential.   In many
nations where agriculture is near a subsistence level, prices
are too low to pay for high technology agriculture.   There is
no strong market demand for food except in bad years.   The
farmer must wait for a disaster to make a profit.   Such an
economic environment will damp-off a green revolution before
it can sprout.
      Finally, some of the charges made against green revolu-
tion in the sociopolitical arena have some validity.   In general, the first to profit are entrepreneurs with access to
enough land to make a good profit and enough capital to avoid
usurious interest rates.   The techniques are not that difficult.   Doctors, lawyers, and politicians can get into the act.
Experience around the world has shown that it is difficult for
green revolution technology to penetrate to the rural destitute.
      In 1971, I joined a group of East African Cereal Workers
for a meeting in Ethiopia and a tour of some of the research

facilities and grain growing regions. One of the tours included the Chilalo Agricultural Development Unit (CADU). This
was a large scale demonstration, financed largely by Swedish
foreign aid. The objective was to install a green revolution
technology on a substantial hectarage in order to show what
the system could do for a country like Ethiopia. High
yielding cultivars were imported along with machinery and
fertilizers. The results were striking. Yields were increased several fold. The increase in wheat production from
this project alone equalled the amount traditionally imported.
In one stroke, Ethiopia had become self-sufficient in wheat
and by extending the system could, it seemed, become an exporter.

But, something looked very strange to me. I know the
Ethiopia landscape very well. On the plateau, at least, it
is full of people. The CADU project looked like Western
Kansas, with fields of wheat as far as the eye could see, but
no one in sight. I asked what had happened to the people. I
was told that the landlords had found the new farming so
profitable that they had evicted the tenants and were doing
the farming themselves with imported tractors. This was, of
course, under the old regime with its largely feudal system.

Several times, I asked: "Where did all the people go?".
I never got an answer. The project was an agronomic success
and a sociologic disaster.

REFERENCE

1. Wellhausen, E. J. *Sci. Amer.* 235, 128 (1976).

INCREASING CEREAL YIELDS:
EVOLUTION UNDER DOMESTICATION

J. M. J, de Wet

Crop Evolution Laboratory
Department of Agronomy
University of Illinois
Urbana, Illinois 61801

I. INTRODUCTION

Cereals are our most important food crops. They are grown on some 70% of all land devoted to food production, and directly supply approximately half of the total food energy consumed every day. In countries such as the United States, Europe and Canada, a large part of the cereal harvest is also processed into secondary food products, or supports the cattle, hog and poultry industries.

Yields are dependent on complex sets of integrated environmental and genetic variables. The simplest way to increase food production is to increase the acreage under cultivation. In Africa, Asia and Latin America the area planted to cereals has increased by some 40% during the last thirty years, and the total yield increased 46% during the same time. Land that can be brought into cultivation, however, is limited, and the best land is already being farmed in all countries. Increases in food production will eventually have to come from increases in yields per acre. This is already true to some extent. During the last thirty years when yield per acre increased by approximately 6% in developing countries, cereal yields per acre nearly doubled in the United States. Better varieties were developed by plant breeders, and farming technology was improved.

The problem of low per acre yields in Africa, Latin America or South Asia, however, cannot be permanently solved by merely transferring high yielding American bred varieties, and American farming technology to these countries where a large part of their food is produced on small individually owned farms. Cereal varieties grown in the United States were bred for very specific habitats, and neither the energy required to prepare such specialized seed beds, nor the funds

to maintain these habitats are usually available to these
farmers. The question is not what the maximum yields are
these farmers can squeeze out of an acre of cultivated land,
but what they can economically produce within their system of
cultivation. Model populations bred for immediate fitness
are high yielding only under essentially ideal environmental
conditions. Environmental fluctuations, under these conditions, can be disasterous to yield. In the United States, an
18% reduction in per acre yield resulted when the maize crop
was attacked by 1970 by a particularly virulent strain of the
pathogen that causes southern cornleaf blight. In this country such a decrease in yield effects the pocket book of the
farmer and consumer. In many other countries, a similar drop
in yield may mean starvation of parts of the population.
Maintaining a basic average yield is more important under
these circumstances than risking total failure. Population
systems of cereal varieties are needed that are relatively
high yielding, but which can at the same time withstand a
range of environmental calamities and still yield reasonably
well. Evolution under domestication produced such population
systems.

II. INCREASING YIELD UNDER DOMESTICATION

        Plant domestication is evolution in man-made habitats.
Wild grasses are commonly harvested as cereals, and some of
these are often sown to increase population size and density.
These practices, however, will not necessarily lead to domestication. American wild rice, *Zizania aquatica* L. has been
harvested for centuries along the shores of lakes in the
northeastern states, but the species retained its wild-type
adaptations (1). Indeed, at least 300 grass species are important wild cereals, but only 33 species are known to have
been regularly cultivated as cereals at one time or another,
and only some two dozen of these have fully domesticated
races.
        Cereals are domesticated when their cultivated races can
no longer compete sucessfully with their wild relatives for
natural habitats. Domesticated taxa are adapted to habitats
created for them by man, and usually also require the help of
man in successful seed dispersal. Some minor cultivated cereals never lost the ability of natural seed dispersal because of selection pressures associated with methods by which
they are harvested (2). Cereals that are harvested at maturity and later thrashed, however, are all characterized by inflorescences in which the spikelets do not disarticulate nat-

urally at maturity. Cereals frequently differ from their
weedy relatives primarily in that the weed is spontaneous in
the man-made habitat, while the crop has to be sown (3, 4).
Weeds, however, may also depend on man for seed dispersal.
The weedy Ethiopian oats (*Avena abyssinica* Hochst. ex Schimp.)
have races with inflorescenes that shatter at time of maturity, while in others the spikelets are persistent. These
latter kinds are harvested with the wheat or barley that they
accompany as weeds, and are regularly sown with the crop (5).

Cereal domestication is initiated as soon as man starts
to sow seeds in specially prepared habitats (6, 7). The domestication process continues as long as seeds harvested from
a sown population are again planted in a cultivated field.

The basic changes in adaptation that accompanied cereal
domestication came about as a result of natural selection
associated with harvesting, sowing and growing in man-prepared
habitats. Cereal evolution was, and is still being accelerated as a result of artificial selection by man to increase
the adaphic range of the species, and to satisfy his own
fancies. Harvesting automatically favors plants with the
highest yield at the time of harvest, and sowing automatically
increases fitness in the man-made habitat. Domestication selects for uniform individual plant and population maturity,
loss of seed dormancy, increase in seedling vigor, and for
yield increase (6).

Domestication emphasizes population fitness. Individuals with the highest reproductive rate are automatically favored. Agronomic fitness is correlated with the number of
progency produced by carriers of selected genotypes relative
to the number of progency produced by other genotypes in the
man-made environment of harvesting and sowing. Differential
reproductive rate under varying environmental conditions promote evolution. When man interferes directly, and selects
for maximum yield under ideal environmental conditions, population variability decreases. Under conditions of less severe artificial selection pressures, sufficient population
variability is maintained to insure progressive evolution
even under extreme environmental fluctuations. Domesticated
races frequently differ very significantly in morphological
traits for their wild ancestors. Grain size is usually increased as a result of selection for increase in seedling vigor. This, however, is not always true. The wild races of
emmer wheat (*T. monococcum* L.) are aggressive colonizers and
resemble the cultivated kinds in seed size. Changes in inflorescence morphology are usually associated with selection
for yield increase. Inflorescences are often greatly enlarged, fertility is restored in sterile florets or spikelets,
number of inflorescence branches is usually increased, and

frequently the overall number of spikelets on each branch is
also increased in number.

Maize (Zea mays L.) and its presumed ancestor teosinte
differ so significantly in gross morphological traits that
they are usually regarded as different species (8). The female inflorescence of teosinte is a distichous spike with solitary fertile female spikelets alternately arranged in indurated cupules that disarticulate individually at time of
maturity (see reference 9). Maize, in contrast, has a highly
specialized female inflorescence. The spikelets are paired,
the cupules are moderately indurated, and the spikelets are
arranged in usually four or more opposite rows around and
along a tough central cob. Yet, maize and teosinte are
genetically conspecific (10), and the characters that distinguish them are phylogenetically of the same kinds as those
that distinguish other domesticated cereals from their closest wild relatives (11-13).

Farmers are traditionally individualistic experimental-
ists. They select genotypes for specific uses, with even
adjacent villages or families often using different kinds of
the same cereal for similar uses. Population variability is
reduced by this practice, but inter-population variability
is increased. Snowden (14, 15), as an example, indicated
that the variability of cultivated grain sorghums (Sorghum
bicolor (L.) Moench) and their closest spontaneous relatives
could be recognized as 13 wild, 7 weedy and 28 domesticated
species. Variability of the domesticated species he further
divided amont 156 varieties and 521 cultivated kinds. These
snowdenian taxa are all conspecific, and associated with
uses by different tribal units or with ecological adaptation
(16-18).

Natural selection increases fitness in the man-made habitats with each generation of sowing. Artificial selection
by man creates differently adapted populations within a range
of variable habitats. Wright (1934) demonstrated that semi-
isolated populations, such as these, provide an ideal system
for rapid evolution. Variability is increased by occasional
hybridization among cultivated races, and hybridization of
these domesticates with related wild races (19). In maize
Mangelsdorf, MacNeish and Galinat (20) demonstrated that its
early evolution was greatly accelerated by introgression of
teosinte, and Wilkes (21) showed that such introgression is
sometimes encouraged by present day farmers where these two
taxa are sympatric in Mexico.

Population variability is essential for progressive evolution. In a cultivated field it protects against severe
fluctuations in yield. Traditional land races are mixtures
of genotypes, well adapted to the norm of environmental vari-

ability, with some genotypes capable of withstanding extreme
environmental fluctuations. They are in equilibrium with
their environment. Yields are low by modern standards, but
consistent. Modern farming techniques demand population
uniformity and maximum yield. Variability is sacrificed for
immediate fitness and maximum yield. These races are better
adapted than traditional ones under ideal conditions. However, their range of environmental tolerance is limited, and
even in the best agricultural regions of the world climate
and other environmental factors do fluctuate. As examples,
a moderately severe drought in South Africa during 1972
caused a 25% drop in average maize yields, and southern corn
leaf-blight reduced American maize yields by some 18% in 1970.

III.   BREEDING FOR YIELD INCREASE

        The potential per acre yield of cereals is tremendous.
The average yield of corn in Mexico is around fourteen bushels per acre. Yield is perhaps fifty bushels in some maize
producing areas but well below the average where conditions
for maize production are not favorable. Fluctuations of
yield within different regions, however, are relatively narrow from year to year. Yield in land races, under traditional agricultural systems, are controlled primarily by genotype
variability within the population. Average yield for maize
in the United States was 97 bushels per acre in 1974. In the
same year the top yield was 230 bushels per acre, and an acre
of maize has yielded 306 bushels under ideal conditions.
Yield in highly derived races is determined by immediate fitness, and the difference between top and record yields is
primarily a function of differences in habitat preparation
and other environmental factors.
        Sorghum, rice and possible other cereals seem capable of
producing some 20,000 lbs. of grain per acre. Top and record
yields are still far above average yields (Table 1). The
energy required to close this gap, by changing the environment, however, is rapidly becoming so expensive, even in the
United States, that this route will soon be unprofitable.
Average yields per acre of cereals are increasing by 2% each
year in the United States, and yield limits seem nowhere in
sight. Since hybrid corn was first introduced, some forty
years ago, average yield has nearly doubled twice as a result of breeding for better adaptation to specific environments. Hybrid sorghums introduced in the middle nineteen
fifties resulted in an immediate jump of nearly 50% in average yield per acre, and consistent progress in increasing
yield has been made since.

TABLE 1

*Average and top yields in the United States, and record yields
in bushels per acre for the major cereals in 1974*

| | | Yield | |
Crop	Record	Top	Average
Rice	350	130	28
Sorghum	320	200	61
Maize	306	230	97
Oats	296	150	51
Wheat	216	135	34
Barley	212	150	43

Success in plant breeding depends to a large degree on
(a) availability of a sufficiently large gene pool, (b) ability to expose desirable alleles through selection, and (c)
techniques to combine selected genotypes into highly adaptive
cultivars with outstanding agronomic fitness. The high
yielding varieties of rice and wheat, developed during the
last decade, have contributed as much toward feeding the
world as hybrid maize and sorghum did when they were introduced into cultivation. Semi-dwarf varieties were developed
that respond favorably to better farming techniques, particularly an increase in available nitrogen. The spread of
these high-yielding varieties across Asia was phenomenal,
from experimental acreages in 1965 to some 40 million acres
in 1975 (22). Under ideal conditions of moisture and available fertilizer these varieties far out-yield standard races,
and they form the basis of the so-called green revolution.
Under adverse conditions, however, growing these specialized
varieties can lead to disasterous crop failures.
    Miracle varieties are rapidly replacing traditional
races across the world's farming acres. They are needed, and
without them starvation cannot be avoided at the present rate
of population growth. They unfortunately also are inherently
susceptible to adverse environmental conditions. It is estimated that at least 70% of midwestern maize hybrids have
two or more of only seven 'station lines' in their pedigrees,
and that these lines were derived from two basic gene-pools,
Lancaster and stiff-stalk synthetic. The 1970 epidemic of
southern corn leaf-blight in the midwest was due largely to

the susceptibility of T-cytoplasm maize to a particularly
virulent strain of race T of the pathogen. A ready solution
was at hand. Returning to normal cytoplasm, maize resistance
was regained. The next epidemic, however, may be more difficult to control, and programs are now underway to increase
the genetic base of maize and other cereals.

Variability has selective advantages in nature, and is
an absolute requirement in successful plant breeding. Should
we reduce variability of cereals to below a critical level on
a world wide scale, we are doomed to eventual failure in our
efforts to feed the ever increasing world population. Increasing yield is evolution under domestication. Evolution is a
change in gene frequencies in populations that are adjusting
to changing environments. Maintaining a reserve of variability for the major crops should be, and is becoming, an
important part of plant breeding projects.

IV.   REFERENCES

1.   Dore, W. G. "Wild rice." Canada Dept. Agric. Publ.
        1393, Ottawa, 1969.
2.   Wilke, P. J., Bettinger, R., King, T. F., and O'Connell,
        J. F., Antiquity 46, 203 (1972).
3.   Harlan, J. R. and de Wet, J. M. J. *Econ. Bot.* 19, 16
        (1965).
4.   de Wet, J. M. J. and Harlan, J. R. *Econ. Bot.* 29, 99.
        (1975).
5.   Rajhathy, T. and Thomas, H. *Miscell. Publ. Genet. Soc.*
        2, 1 (1974).
6.   Harlan, J. R., de Wet, J. M. J. and Price, E. G.
        Evol. 27, 311 (1973).
7.   de Wet, J. M. J., *Bull. Torrey Bot. Club* 102, 307 (1975).
8.   Wilkes H. G., "Teosinte: The closest relative of maize.
        Bussey Inst. Publ.", Harvard Univ., Cambridge, Massa-
        chusetts, 1967.
9.   Mangelsdorf, P. C. "Corn, its origin, evolution and
        improvement" Belknap, Harvard Univ. Press, Cambridge,
        Massachusetts, 1974.
10.   Collins, G. N. and Kempton, J. H., *J. Agric. Res.* 19, 1
        (1920).
11.   Galinat, W. C., *Ann. Rev. Genet.* 5, 447 (1971).
12.   Beadle, G. W., *Field Mus. Nat. Hist. Bull.* 43, 2 (1972).
13.   de Wet, J. M. J. and Harlan, J. R., *Euphytica* 21, 271
        (1972).
14.   Snowden, J. D. "The cultivated races of sorghum"
        Ylard and Son, London, 1936.
15.   Snowden, J. D., *J. Linn. Soc. London* 55, 191 (1955).

16. de Wet, J. M. J. and Harlan, J. R., *Econ. Bot.* 25, 128 (1971).

17. Stemler, A. B. L., Harlan, J. R., and de Wet, J. M. J., Evolutionary history of cultivated sorghum (*Sorghum bicolor* (L.) Moench) of Ethiopia. *Bull. Torrey Bot. Club.* 102, 325 (1975).

18. Stemler, A. B. L., Harlan, J. R., and de Wet, J. M. J., *J. Afr. Hist.* 16, 161 (1975).

19. Harlan, J. R., *Euphytica* 14, 173 (1965).

20. Mangelsdorf, P. C., MacNeish, R. S., and Galinat, W. C. "The Prehistory of the Tehuacan Valley," (D. S. Byers, Ed.). Univ. Texas Press, Austin, Texas, 1976.

21. Wilkes, H. G., *Bot. Mus. Leaflets, Harvard Univ.* 22, 297 (1970).

22. Dalrymple, D. G., *USDA-AID Foreign Agric. Econ. Report,* 95, 1 (1976).

HEVEA RUBBER: PAST AND FUTURE

Ernest P. Imle

Agricultural Research Service
U.S. Department of Agriculture
Hyattsville, Maryland 20782

I. INTRODUCTION

A tree gum used by natives to make bouncing balls caught the attention of the first European visitors to the New World. They took samples of this gum back to Europe where the English, upon discovering that it could be used to rub out marks made on paper, called it rubber. It still bears this name in all English-speaking countries. From this humble beginning as a new world curiosity, natural rubber has developed into an important crop for the lowland tropics. About 3.4 million metric tons of it are produced annually and its uses are legion. I shall mention some of the events which made it possible for rubber to reach its present importance in agriculture, industry and world trade.

Discovery of vulcanization of 1839 by the American chemist, Charles Goodyear, was the first big step toward improving the properties of crude rubber. Goodyear, whose memory is honored in the name of one of the major rubber companies, heated rubber with sulfur and produced a product that was vastly superior to the raw material. His vulcanized rubber was less sticky than raw rubber, had greater strength and more resistance to heat and cold, and was altogether more suitable for a multiplicity of purposes. Goodyear's landmark discovery was soon followed by innovations in the fields of solvents, additives, compounding, processing, molding and curing.

When the auto age began to roll on pneumatic tires, it was clear that none of the wild tree sources would be able to meet the foreseeable demands for rubber. Rubbertree culture somehow had to be developed and this brings us to the story of *Hevea brasiliensis* (Willd. ex A. Juss.) Muell. Arg., commonly known as the Brazilian rubbertree. It is native to the Amazon River basin and belongs in the family *Euphorbiaceae*. The

119

superior qualities of the natural elastomers produced by this
tree have never been surpassed by any of the synthetic products.

Before proceeding with the story of *Hevea*, however, I
wish to mention some of the other rubber-producing plants
which have played a part in the story of rubber. The rubber
balls which Columbus saw the Indians using in Central America
and the Caribbean almost certainly were made from the latex of
trees of one of the species in the genus *Castilla*, which
belongs in the *Moraceae* or mulberry family. The rubber balls
seen by Cortez in Mexico may have been made either from
Castilla or from the desert shrub known as guayule. Several
of the *Castilla* species produce much more latex at a single
tapping than any *Hevea* tree will yield. However, they will
not produce such yields if tapped more frequently than once or
twice per year. *Castilla* rubber is of good quality and wild
trees of this genus contributed substantially to U.S. rubber
supplies in World War II when every available wild tree was
sought out and tapped. *Castilla* could almost surely be greatly improved as a rubber producer through research. Wild trees
of this genus still supply rubber for local rubber needs in
the tropics (Fig. 1). Guayule, *Manihot*, *Funtumia*,
Cryptostegia, Russian dandelion and many other plants have
been exploited for rubber to some extent in the past, especially in times of high prices, but in the long run none could
compete with *Hevea*. Interest in guayule rubber has waxed and
waned over the years and there has been a continuing production of it right up to the present day. Guayule culture is
once again attracting attention because its competitive position has improved with the rise in cost of raw materials needed for making synthetic rubber, and it can grow and produce
rubber under lower rainfall conditions than many other crops.
Several of the above-mentioned, rubber-producing plants have
an interesting history, but this paper will deal only with
some highlights in the story of *Hevea*.

II.   HEVEA MOVES TO ASIA

It is fitting that there should be a paper on *Hevea* in
this symposium in 1976. It was in 1876, exactly 100 years ago,
that Henry A. Wickham collected several basketfuls of seeds of
Hevea brasiliensis in Brazil at a site along the Tapajoz River.
Wickham succeeded in transporting these delicate, short-lived
seeds across dense forest to Santarem and down the Amazon
River 640 km to Belem. From there they were shipped across
the Atlantic to the Kew Botanical Gardens in England. Some
seeds still were viable upon arrival at Kew and the records
state that approximately 2700 seedlings were obtained.

Two thousand of the best seedlings were shipped in

Fig. 1. Tree of Castilla elastica in Nicaragua showing
marks of recent excessive and destructive tapping; even the
exposed roots have been tapped.

Wardian cases from Kew to botanic gardens in Ceylon, Malaya
and Java. Some reports state that the bulk of them went to
Ceylon and that only 22 seedlings out of the total shipment
reached Malaya. In any case, we know that the rubbertrees in
southeast Asia, millions of hectares of them, are derived from
a very few plants of Wickham's original stock from the banks
of the Tapajoz. A few seedlings from the original lot were
sent from Kew to botanic gardens in other tropical locations
and some of these original trees still survive. I have been
privileged to see and study one of them, now a magnificant
tree that flourishes near the fern house in the Botanical
Garden of Queen's Park in Port-of-Spain, Trinidad. Botanic
gardens played a big role in establishing Hevea as a new crop
just as they have with a number of other new agricultural

enterprises.

Several of the circumstances and outright accidents relating to the Wickham seeds will be of special interest to
economic botanists. Wickham did not know it then, but is now
known that *H. brasiliensis* is the only *Hevea* species indigenous to the Tapajoz area where he collected. Had he collected
elsewhere, he probably would have come up with seeds of one of
the other less productive *Hevea* species because he could not
at that time have known enough about the genus to select only
H. brasiliensis. Had he collected one of the other *Hevea*
species, the development of rubber cultivation would have been
delayed everywhere, perhaps by many decades.

In two other respects, however, Wickham's choice of collection site was less fortunate: 1) Yield potential of the
Tapajoz populations of *H. brasiliensis* is lower than that of
the populations found farther up the Amazon. This fact, not
discovered and demonstrated until the 1940's, probably delayed
the development of high-yielding trees by plant breeders in
southeast Asia. 2) The Tapajoz population was all highly susceptible to South American Leaf Blight (SALB), the disease that
to this day has thwarted, delayed or hampered all attempts to
develop rubber plantings in the Western Hemisphere. This disease, caused by the fungus *Microcyclus ulei* (P. Hemm.) Arx,
was not recognized in Wickham's time. Polhamus (1) states in
his book, *Rubber*, that it was clearly a stroke of luck that
Microcyclus was not carried along with the seeds to Kew and
thence to the Orient on the seedlings. If it had been, we can
be sure that rubber planting in southeast Asia could not have
succeeded and the story of rubber would have been a very different one. To this day, SALB has not invaded the rubber
plantings of either southeast Asia or Africa. When and if it
does, it will surely cause major economic dislocations for the
millions of pwople associated with rubber production.

Successful transfer of the Wickham material from South
America to the Orient, leaving SALB behind, was certainly the
first big step in domestication of *Hevea*. Arrival of some of
the Wickham seedlings at the Singapore Botanic Garden in
Malaya set the stage for the second and equally important step,
namely the development of a superior tapping system by Mr. H.
N. Ridley. Ridley saw that the haphazard and destructive
tapping methods used on wild trees would not do for a permanent plantation crop. After patient and brilliant work, he
came up with an improved tapping system which caused minimal
tree damage. It economized on bark consumption, allowed a
tree to be tapped 100 or more times per year and greatly
increased annual yields. His tapping method, with minor modifications, remains in use today (Fig. 2).

Fig. 2. *Hevea brasiliensis seedling tree being tapped by a variation of Ridley's system. The previous cut is reopened every third day by removing a thin strip of bark. The cut is made as deep as possible without actually wounding the cambium. Under conditions of proper tapping, the tree renews the exploited bark and it becomes available in due time for retapping.*

A recent successful innovation in the Ridley system is to tap upwards on the panel rather than downwards. Half-spiral cuts opened every fourth day on the upward-cut system are reported to be giving as good yields as half-spiral cuts opened every third day on the downward-cut system.

III. GENETIC IMPROVEMENT

Until the early 1900's, there still was disagreement as to whether *Castilla*, *Hevea*, *Manihot* or some other plant should be chosen for cultivation on a large scale. The southeast Asian experience with the Wickham progenies soon made it clear,

however, that *Hevea* was the best choice. It also became clear
that yields per tree and per hectare would have to be improved.
British, Dutch, French, and local specialists, and later North
and South American scientists as well, all had a part in developing better planting stocks and improving culture of the
trees. The first unselected Wickham trees had a yield potential no higher than 225 kg per hectare per year, even with
Ridley's best tapping methods. Crosses between some of these
original trees gave encouragement to the early workers in the
form of vigorous progenies that out-yielded their parents. It
has been suggested that a number, perhaps all, of the original
Tapajoz seedlings were in fact inbreds, as a result of many
generations of natural selfing of individual, isolated, wild
forest trees, and that subsequent crosses between seedlings
from some of these natural inbreds produced progenies which
exhibited heterosis. These early-generation, vigorous seedlings offered promising stocks from which additional superior,
higher-yielding trees and clones eventually were derived.

Budgrafting of *Hevea*, developed in Java by Dutch horti-
culturists, was the third great step in domestication of this
plant. Budgrafting made it possible to propagate quickly and
to the extent desired any superior seedling selection and to
establish plantations of high-yielding clonal trees. Without
a successful method for vegetative propagation, it would not
have been possible to use genetically improved trees on a
large scale. Over the years improvements in budgrafting have
been made, including its adaptation to very young seedlings
and to crown budding, but the basic patch-budding method used
is the same as that developed in Java.

Average plantation yields had risen above 400 kg per
hectare by the 1930's and some experimental blocks of clonal
rubber by that time were yielding approximately 1000 kg per
hectare. More recent improvements have raised average commercial plantation yields to 1200 kg per hectare and some plantations do better than 1600 kg. Planting materials with a potential above 3000 kg per hectare now are available (Fig. 3).
In general, large plantations have better management and higher yields than the small holders' plantings. Yields of the
latter tend to vary greatly, but there are many examples of
small holder yields equalling those of the best plantations.
In summary, we can note that productivity of the *Hevea* tree
has been increased more than tenfold above that of its wild
Tapajoz River progenitors. We believe few other crops can
equal this record over the last 100 years.

IV. HEVEA REBOUNDS TO TROPICAL AMERICA AND TO AFRICA

In the 1920's, high prices and threats of monopoly spur-
red new attempts by U.S. interests at plantations in South

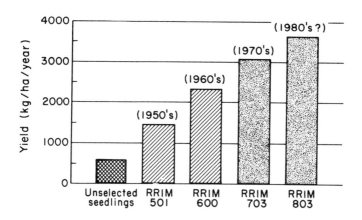

Fig. 3. *Progress in breeding and selection for yield at the Rubber Research Institute of Malaya (from Planters Bulletin, November 1975).*

America and in West Africa. Plantings in Liberia, started by Firestone, prospered on the lateritic soils of that country and have contributed heavily to world rubber needs and to the Liberian economy. Production in Africa eventually spread from the large company plantations onto private, native holdings just as it did in southeast Asia. Today, approximately half the total production in both areas comes from small independent holders. Research sponsored by company plantations and by governments continues to provide improved plant materials and technology while the private growers today do most of the actual planting and production.

Attempts at planting in South and Central America did not go as well as those in Africa. The Ford Motor Company's plantations in Brazil, started in 1928, and those of Goodyear in Costa Rica and Panama, started in the 1930's, ended up as costly failures because of SALB. These companies were defeated in trying to plant the genetically improved stocks from southeast Asia because all of this material was highly susceptible to SALB. Furthermore, they quickly found that all the accessible Western Hemisphere material not only was low-yielding, but almost all was likewise highly susceptible to SALB. It was evident after a time that, unless special ecological zones unfavorable to SALB could be found in the Western Hemisphere, the only solution lay in a breeding program to combine

high yield with SALB resistance. Even though these early and
costly private company efforts produced little or no rubber,
they were not in vain. They provided a base for a cooperative
rubber research program among the governments of Brazil and
other tropical American countries, the U.S. Department of Agriculture (USDA) and private rubber companies. This program,
which started in 1940 and continued well into the 1950's, will
be described in some detail because it opened important new
chapters in the story of *Hevea*.

V.   THE RUBBER CRISIS OF WORLD WAR II

It is hard for today's younger scientists to appreciate
the critical situation our country faced in 1939 with regard
to rubber. The war in Europe was enlarging, access to southeast Asian rubber was in peril, wild rubber production had almost ceased, stockpiles were inadequate, synthetic rubber
capacity was insignificant, quality of known synthetics was
inferior for most critical uses, and the necessary research
had not been done to ensure successful establishment of Western Hemisphere plantings. This crisis sparked three programs
that have greatly affected the story of *Hevea*: 1) The synthetic rubber program, 2) The Rubber Development Corporation set
up within the U.S. Department of Commerce to promote wild rubber collections, and 3) The Cooperative Rubber Research Program which was led by the USDA.

1) The U.S. government's synthetic rubber program was an
enormous success. It still is regarded as one of our country's
major industrial achievements. Synthetic rubbers, poor in
quality at first, were soon improved and special types were
developed. The superior rubber molecule produced by *Hevea* is
chemically *cis*-1,4-polyisoprene. It has a high degree of geo-
metrical regularity, a molecular weight above 1 million and is
not easily duplicated synthetically. However, one type of
synthetic rubber, synthetic polyisoprene, is an approximate
replica of the *Hevea* molecule. The synthetics soon relieved
the country from utter dependence on the dwindling stockpiles
of natural rubber (NR) and on the meager supplies of wild
rubber. By the end of the war, synthetic rubber (SR) was well
established as an acceptable raw product. In fact, by this
time, SR had several advantages over NR. The SR types could
be designed for special industrial uses and supplied in standard uniform quality. Also, SR was not subject to undertain-
ties of long and costly ocean shipments. Rapid capture by the
synthetics in this period of a large part of the booming rubber market led some economists and others to predict the early
decline of tree rubber as a world commodity. Events have dis-
proved this prediction in that production of NR has in fact

tripled since World War II. Over 3.4 million metric tons of
NR now are used annually, about one-third of total world consumption (Tables 1 and 2). NR has retained supremacy for some

TABLE 1

Basic Rubber Statistics, 1970
(from Malaysian Rubber Research and Development Board)

Region	Rubber consumption ('000 tons)	NR consumption ('000 tons)	NR share (%)
U.S.A.	2517	568	22.6
EEC	1859	701	37.7
Japan	779	283	36.3
Subtotal	5155	1552	30.1
Analysis total	3857	1240	
Free World Total	6755	2290	33.9
Centrally-planned countries total	(1845)*	700	(38)
World	(8600)	2990	(35)

*FAO estimate

of the most demanding uses, such as airplane tires, which
still contain 90 to 100 percent NR (Table 3). Heavy duty
equipment tires, for rugged use in off-the-road conditions,
also require natural rubber because there is less destructive
heat buildup with NR.
    Botanists will be amused by the following definition
which used to be jokingly applied to some of the earlier, far
from satisfactory, SR types: "Synthetic rubber is a wonderful
material. It can be mixed in any proportion with natural rubber and the more natural one adds the better it is." Today it
is not a question of NR *versus* SR, but one of NR and SR since
both are needed. They are somewhat complementary and each
also has certain preferential uses.
    2) The Rubber Development Corporation (RDC) was set up to
increase wild rubber production from all sources, almost regardless of cost, in order to augment dwindling NR stockpiles.
This program was a successful cooperative effort between the
United States and the tropical American countries. The valuable contributions of our tropical American neighbors in this
work is well remembered by all who participated. After the

TABLE 2

Consumption and Market Share 1900-1970
(from Malaysian Rubber Research and Development Board)

Year	Total World rubber consumption ('000 tons)	NR consumption ('000 tons)	NR share of the market (%)
1900	53	53	100
1910	102	102	100
1920	302	302	100
1930	722	722	100
1940	1127	1127	100
1950	2339	1750	75
1960	4400*	2095	48
1970	8600*	2992	35

*FAO estimates, including SR consumed in centrally planned countries.

TABLE 3

Natural Rubber Share of Each End Use, Percent, 1970
(from Malaysian Rubber Research and Development Board)

Product		U.S.A.	EEC	Japan[a]
Types:	car	12.5	22	28
	truck/bus	55	75	50
	tractor, etc.	30	60	n.a.
	bicycle/m'cycle	10	30	63
	aircraft	90	90	100
	retreading	12	33	54
	inner tubes	5	5	10
Latex products		22	33	n.a.
Belting)		25	30	52
Hose)			25	43
Footweat		25	35	44

[a]The Japanese figures are calculated directly from official statistics, quoted here without rounding off.

war ended, RDC went out of existence, its job having been well done by providing the United States with many thousands of tons of wild rubber it could not otherwise have had. An extra dividend from the RDC work was the opportunity it gave for botanists and rubber specialists to penetrate into remote areas in the Amazon basin to study local *Hevea* populations and to obtain new germplasm for research stations.

3) The USDA cooperative Rubber Research Program was initiated in 1940 when an alarmed Congress decided that dependable sources of rubber had to be developed for use within the Western Hemisphere. USDA scientists who had experience in tropical agriculture, Dr. E. W. Brandes, Dr. R. D. Rands and Mr. L. G. Polhamus, were selected to organize and lead this work. Besides *Hevea*, the program also included guayule, Russian dandelion, golden rod and *Cryptostegia*. For *Hevea*, there was set up a superb, international, tree crop improvement program, the goal being to develop commercial, high-yielding, disease-resistant planting stocks of *Hevea* for SALB-infested areas. A coordinated plan was devised to achieve this goal and despite the constraints of emergency wartime conditions, the program moved ahead. Surveys were made, central testing stations were established, clones and seeds were assembled, bilaterial agreements were signed with 13 countries and plant scientists were hired and deployed to strategic areas.

By 1954, after 14 years of operation under a modest budget, and despite some major setbacks in the form of new and unexpected diseases, impressive progress had been made. Several hundred clones with good levels of resistance to the two major leaf diseases had been developed, some of them with near-commercial yields. Areas in the Amazon basin with superior and useful germplasm had been delineated. Methods for controlling SALB in nurseries had been developed. Advances were made in production and use of the 3-component tree, a tree consisting of a seedling rootstock, a high-yielding trunk clone and a disease-resistant canopy clone united into a single plant through budgrafting. These 3-part trees, the product of what we could call horticultural engineering, provided interim planting stocks, pending success with the longer-term genetic engineering approach. A full description of this program and its achievements is given by Rands and Polhamus (2). Discovery and development of resistance to SALB under this program can be considered a fourth great step in *Hevea* domestication.

Despite the program's many achievements, it became necessary by 1954 for the USDA to withdraw from it. The rubber crisis had ended and alternate uses for rubber research funds became more attractive. However, private companies and other agencies continued to make use of the experience, results and germplasm that had been accumulated. In the absence of a single coordinated program it has taken longer to approach the goal but planting stocks are now available for use in some parts of the SALB-infested Western Hemisphere tropics. The ongoing research is expected to produce still better ones. Ultimately stocks with a horizontal type of resistance to SALB may be produced.

It was well understood in some quarters that commercial planting stocks with resistance to SALB would have enormous future importance to Asia and Africa. The vast rubber areas there had been shown to be entirely susceptible to SALB. Resistant clones could provide some insurance and protection if SALB should spread to those areas. Arrangements were made, therefore, for some of the newly selected plant materials to be introduced into southeast Asia and later into Africa. This was the first significant infusion of improved *Hevea* germplasm since the arrival of the original Wickham seedlings in 1877.

VI.   RUBBER COMPANY RESEARCH AND DEVELOPMENT

The Goodyear Company conducted testing and development work in Guatemala and also opened a plantation near Belem, Brazil. They have continued to the present date working toward blight-resistant, commercial planting stocks using mainly special *Hevea Benthamiana* Muell. Arg. selections as sources of genus for resistance. However, they still find it necessary to use aerial spraying on their Brazilian plantation during the annual leaf-change period in order to control SALB and get good annual refoliation.

The Firestone Company, under the guidance of their Research Director, Dr. K. G. McIndoe, went on to develop an admirable program which has continued to the present time. They have concentrated on germplasm from the Madre de Dios area of Peru where USDA scientists had found wild *Hevea* trees populations with high quality rubber, higher than average yield, and a notable level of resistance to SALB. Using duo-clone breeding gardens and isolated seed fields on their plantations in Guatemala and Liberia, Dr. McIndoe conducted 15 years of rubbertree breeding for Firestone, with high-yielding but susceptible clones from southeast Asia and resistant selections from the Madre de Dios area. Promising seedlings were tested for resistance, cloned and divided between Guatemala and Liberia, using an intermediate station in Florida for shipments in each direction. Over 2500 clones have been transferred to date. The best are now in yield tests and some commercially suitable, blight-resistant ones have been identified. Firestone is also incorporating germplasm of *H. pauciflora* (Spruce ex Benth.) Muell. Arg. into this breeding program as a means of broadening the genetic base for resistance to SALB. Some of the hybrids between *H. brasiliensis* and *H. pauciflora* have shown remarkable vigor and growth rates in tests on some of the poor soils of the Amazon basin (Fig. 4).

Fig. 4. Center row, seedlings from H. pauciflora X H. brasiliensis. On each side, seedlings of brasiliensis X brasiliensis parents. All of equal age growing on poor soils at Belterra on the Tapajoz River near Santarem, Brazil.

Since 1954 Firestone has developed in Bahia, Brazil, in an SALB-infested area, a highly successful 6000 hectare plantation using *Hevea* selections which have some degree of resistance to SALB. Their production is approximately 1200 kg per hectare. They plan to augment the Guatemala-Liberia breeding project described above by doing a large amount of the future

research and selection work on their Brazil plantation. By
continuing this well-conceived, interhemispheric, rubber
breeding project, Firestone will help to insure the future
availability of safe planting stocks for use of small farmers
in the developing countries of tropical America and Africa.
Benefits will accrue also to the Asian small farmer in the
form of insurance against SALB.

In 1954, the B. F. Goodrich Company, the only one of the
big four U.S. rubber companies (Goodyear, Firestone, Goodrich
and U.S. Rubber Co.) that had never engaged in natural rubber
production, started a new plantation in Liberia. This change
in course on the part of Goodrich was an interesting indication that SR was not a complete answer to industry's rubber
needs. They developed a 8,000 hectare plantation in a part of
Liberia well removed from the older Firestone plantings. This
new Goodrich project benefited by having as its technical director, Mr. William MacKinnon, who had had a long career in
rubber in southeast Asia and also had served in the new world
tropics under the Rubber Development Corporation and the USDA
cooperative rubber program. He brought to the project knowledge and technology from all these sources.

Several other rubber companies have made recent and significant investments in rubber plantation development. Among
these are Uniroyal, Michelin and Pirelli. Uniroyal, formerly
called the U.S. Rubber Company, had started plantings in
Sumatra as long ago as 1910 and later in Malaysia. In 1965,
they began new developments in west Africa. Michelin is put-
ting in new plantations in the Ivory Coast Republic. Pirelli
has a plantation development near Belem in the state of Para,
Brazil. While these developments have not included a strong
research element on SALB, similar to that made by Firestone,
they have served to transfer advanced rubber technology into
new areas.

The Brazilian government continued after 1954 with some
parts of the original rubber program, and they have cooperated
with U.S. and European companies in establishing tests and
plantations in the Amazon and in Bahia. The Rubber Research
Institute of Malaya (RRIM) is cooperating with the Brazilian
government in continuing exchanges of rubber germplasm, using
a station in Trinidad for resistance screening. The RRIM is
concentrating on sources of SALB resistance from *H.
brasiliensis* populations other than the Madre de Dios material
and also using *H. Benthamiana* and *H. pauciflora* as sources of
resistance and vigor.

VII. FUTURE OF HEVEA

Predictions that NR is on the way out are not heard any-
more. World markets take every ton of natural rubber produced.

Price trends indicate that they would take more if it were
available. It is estimated that consumption in the early
1980's may be 16 million metric tons with natural rubber supplying 37 percent of this, or 6 million metric tons. The
world's present rubber plantings cannot supply 6 million
metric tons. However, by replanting older areas with superior
clones, plus extensive use of yield stimulants, plus some new
plantings, a 6 million metric ton annual crop may be possible
by the mid-1980's. Much of the land now devoted to rubber is
not highly regarded for food-crop production so no great di-
version of land from rubber to food crops is likely to take
place in the near future, but there will be trends in this
direction as human populations continue to rise. This can be
countered in part by intercropping young rubber with food
crops, a practice which, when properly controlled, can actually favor growth of the young rubber.

Production of the synthetic rubbers appears to have reached a point at which costs increase almost directly with increases in wage rates in the industrialized countries. This,
coupled with the rising costs of the petroleum-based raw materials needed for manufacture of synthetics, can be expected
to put NR in a better cost-competitive position than at any
time in the last 25 years. Furthermore, there are some cost-
reducing adaptations still to be widely employed in NR production, such as improved collection and processing of crude lump
rubber and latex from small holders, and the more extensive
use by small planters of higher-yielding new clones.

The strong trend toward use of radial tires is favoring
greater use of NR. For best performance radials require a
higher percentage of NR than has commonly been used in passen-
ger vehicle tires since World War II.

Under conditions developing in the world today, it
appears to me that *Hevea* as a crop for some of the developing
countries holds more promise than at any time in the past. As
a tree crop, *Hevea* causes a minimum of damage to the environment and little, if any, soil degradation, in contrast with
annual cropping and the slash and burn system, both of which
have well-known destructive effects on soils and environment
in most parts of the humid tropics. The crops is well suited
to the small or independent land holder. It is labor intensive and could help reduce the disastrous migration from rural
areas into city slums caused by a lack of employment opportunities in rural tropical areas. Rubber tapping is best done
in the early daylight hours. This leaves time for production
of food crops and other farming activities. Rubber is easily
stored and is a readily salable export product for which the
market is increasing. At the same time it is a prime material
on which to base industrialization in the developing countries.
A rubber plantation can produce rubber for 16 to 30 years be-

fore it has to be replanted. By that time, it will have a
biomass consisting of 250 to 375 large trees per hectare which
can be used for fuel, chipboard, pulp, or rough lumber. Proper use of this residue may cover costs of replanting a plantation with improved clones.

Expansion of *Hevea* culture in the Ivory Coast Republic of
Africa is receiving expert guidance through cooperative efforts
between French and Ivory Coast scientists and plantation ex-
perts. The best kept and most impressive young *Hevea* plantations I have ever seen were in the Ivory Coast a few years ago.
Continuation of these high standards of development plus the
scientific cooperation and backstopping from the excellent
French rubber research groups will soon make *Hevea* an important crop in the Ivory Coast.

The Amazon basin, home of *Hevea*, is often mentioned as a
logical place for extensive plantings. However, the disease
problems are formidable and it is known that extensive soil
areas where *Hevea* grows wild are not suitable for plantation
rubber. Another problem in some areas will be the early morn-
ing rains which will interfere with tapping. Although *Hevea*
occurs naturally throughout much of the Amazon basin, only one
or two wild trees are found per hectare among a great mixture
of other tree species. SALB does little damage of these
scattered *Hevea* trees, but it becomes devastating when *Hevea*
monoculture is attempted. Barring the unexpected discovery of
a practical chemical control for SALB, *Hevea* culture in the
Amazon basin cannot succeed unless it is based on planting materials with multiple-gene type of resistance, or a horizontal-
type resistance, to major strains of *Microcyclus ulei*. True
immunity to SALB probably never will be found, but plantings
with a canopy of clones that carry resistance factors of diverse origins can be expected to produce well in spite of SALB.
Local testing of all new material will be required before it
can be safely used in any area. Recent reports from Brazil
tell of that country's plan to establish a large rubber research center at Manaos, in the heart of the Amazon basin.
Such a center may show the way to use selected parts of the
basin for profitable rubber production and help to restore
this product to an important position in the Amazonian economy.

Many special plant-derived products are synthesized or
even improved upon by the chemist soon after being discovered
and thereafter are produced in a factory rather than by agriculture. *Hevea* rubber, as we have pointed out, has held its
own against the best efforts of the chemists. It has remained
economically competitive mainly because some of the NR producing countries maintained strong support for research and
technology. RRIM, the world's largest single crop research
institute, has been the leader in this successful fight to
maintain *Hevea* rubber's strong position. The RRIM has made

striking advances in breeding for yields (Table 1). They now
consider 5000 kg per hectare as a possible goal. Another notable contribution of the RRIM has been in development of standardized, special-purpose type of NR to meet specific industry
needs. The RRIM also has been a leader in research on the use
of chemical yield stimulants on tapping panels. These stimulants can induce yield increases of 25 percent on an annual
basis. Additional research on this subject will bring more
valuable results. Yield stimulants could be considered as the
fifth great step in *Hevea* domestication. However, all clones
do not respond equally well to use of yield stimulants. Im-
proper use of stimulants can result in a reduction in yields
at some future date and in poor tree growth.

VIII. RESEARCH NEEDS

High grade intensive research must continue and its application must be expedited if 6 million metric tons of *Hevea*
rubber are to be available by the mid-1980's. There persists
a wide gap between potential yields, as determined in experimental blocks, and the actual yields obtained in the plantation or field, even with good management. Clones and planting
stocks are needed that are less environment-specific so the
average small farmer can benefit from higher yields. A research area that holds promise for increasing yields but which
has not received serious attention is the development of special rootstocks for clonal rubber. Research on this subject
might lead to a sixth great step in rubber domestication and
improvement.

Even though more new germplasm has become available to
Hevea breeders in recent years than in the whole previous history of the crop, there is a need for germplasm with more diverse sources of resistance to SALB. There also is a great
need to locate, assess and somehow preserve additional material. A number of *Hevea* species in the Amazon have been studied inadequately or not at all for their potential in breeding
for vigor, disease resistance and yield, or for their possible
use as rootstocks, or for developing special hybrids which
could be used to advantage as rootstocks. There are several
genera closely related to *Hevea* which may have value in genetic improvement programs but which have not been assessed. It
is reported that a long-term study of Amazonian flora is soon
to be initiated by the National Research Council (C.N.Pq) of
Brazil and that U.S. participation has been invited. This
study should provide opportunities to locate and collect or
preserve additional germplasm within the family *Euphorbiaceae*.

An additional warning note has to sounded about SALB. If
this disease somehow reaches southeast Asia or Africa, it
surely will reduce the world's supplies of *Hevea* rubber. For-

tunately, the governments and rubber interests, particularly in Malaya, are fully alert to this possible disaster and are prepared to attempt eradication if the disease should appear. Meanwhile, there is no time to spare in developing and using a good backlog of clones, cultivars and planting stocks with broadly based resistance to strains of the causal organism, *Microcyclus ulei*.

We have pointed out that the story of *Hevea* as a crop covers barely a century, and we have mentioned five great steps in *Hevea* domestication.

1) *Hevea* transfer by Wickham from South America to the Orient, leaving SALB behind.

2) Development of a superior tapping method by Ridley.

3) Vegetative propagation by budgrafting.

4) Discovery and development of *Hevea* selections with resistance to SALB.

5) Use of chemical yield stimulants on the tapping panel. The stage appears to be set for an eventful second century in which the crop will play an increasingly important role in the developing countries.

For those who wish to delve more deeply into the botanical, agronomic and technical story of natural rubber we recommend the book, *Rubber*, by Polhamus (1). For the romantic and exciting but also tragic early history of rubber, the reader should turn to the book by Vicki Baum entitled, *Weeping Wood* (3). An interesting account of this tree in Liberia, which covers its improvement through an international exchange and breeding program, is to be found in a booklet, *The Rubber Tree in Liberia*, by K. G. McIndoe (4).

IX. REFERENCES

1. Polhamus, L. G., "Rubber, Botany Cultivation and Utilization" Interscience Publishers, Inc., New York, 1962.
2. Rands, R. D., and Polhamus, L. G., U.S. Dept. Agric. Circular No. 976, June 1955.
3. Baum, V., "The Weeping Wood" Greenwood Press, Inc., 1971.
4. McIndoe, K. G., "The Rubber Tree in Liberia" John McIndoe Limited, Dunedin, New Zealand, 1968.

HORSERADISH - PROBLEMS AND RESEARCH IN ILLINOIS

A. M. Rhodes

Department of Horticulture
University of Illinois
Urbana, Illinois 61801

I. INTRODUCTION

Horseradish, *Armoracia rusticana* Gaertn., Mey. & Scherb.,
a perennial of the Cruciferae, is grown for its pungent roots.
It is thought to be indigenous to temperate Eastern Europe
and probably has been in cultivation for less than 2,000 years
(1). The plant forms a rosette of erect, long-petioled
leaves. The roots are fleshy, corky-tan externally, white
within, conical at the top, and abruptly branched below to
form a deep root system which is difficult to eradicate. The
inflorescence bears a mass of small white flowers in panicu-
late racemes. Pods contain 1 to 6 seeds. Although the plant
was reported to be highly sterile by early botanists, present
genotypes are predominantly fertile (2).
    About half of the 6 million kg of annual commercial pro-
duction in the United States is grown in the fertile soils of
the Mississippi River Valley around East St. Louis, Illinois.
Only 2 other areas, Eau Claire, Wisconsin and Tulelake, California have sizable acreage. Total U.S. commercial acreage is
estimated at 800 ha. Prices received by growers for crude
horseradish fluctuate widely from year to year. When there
is a short supply of the domestic product, prices increase
and horseradish is imported to supplement the domestic supply
(3).
    This report deals mainly with some of the past and pre-
sent problems confronting Illinois horseradish growers and
the results of research at the University of Illinois directed
towards solving these problems. Much of this information has
been presented at grower meetings and has not been published.

II. CULTURE

Horseradish is grown as an annual. A large marketable
root develops from a set or secondary root that increases in
diameter to become the marketable or primary root. One year
old secondary roots are selected from 2 year old primary
roots for planting the following year.

The highly specialized culture of horseradish requires
much skill and labor. The following method of culture is
that generally employed in the East St. Louis area. Set
roots ¼ to ½ inch in diameter and 12 to 14 inches long are
placed obliquely to the horizontal in shallow furrows with
the crown end resting slightly higher than the basal end.
The sets are covered with 2 to 4 inches of soil, followed by
rolling to firm the soil. Spacing is 18 to 24 inches within
the row and 36 to 40 inches between rows. After the plant is
well established, the crown end of the set is lifted slightly
causing the new secondary roots on the upper portion to break
-off. The secondary roots of the basal end nourish the set
and young plant. If secondary crown roots are permitted to
develop, they tend to prevent enlargement of the set root
resulting in fewer high quality marketable roots.

The roots are either dug in late fall or early spring
using a modified moldboard plow or a specially built mechanical digger which operate to a depth of about 20 inches. At
this depth long secondary roots are obtained for new sets.

After harvest secondary roots are broken off and sets
to be kept for planting the next crop are buried in outdoor
pits or cool cellars. Primary roots are either graded and
sold immediately or stored under cool moist conditions for
sale later.

Culture studies by the Department of Horticulture have
centered around the theme of producing a better quality root.
Much of this work has been done on commercial farms. Studies
have included weed control, root growth, soil fertility,
growth regulators, and variety trials.

Through cooperative work with chemical companies and the
USDA, H. J. Hopen and C. C. Doll (4) helped to obtain clearance of a herbicide, nitrofen (TOK), for control of annual
grasses and purslane, *Portulaca oleracea* L., a troublesome
weed of horseradish. Chemical clearance is an expensive
procedure and for a minor crop, such as horseradish, the cost
return ratio must be carefully considered. An additional
herbicide, DACPA (Dacthal) has been granted a tolerance for
use in horseradish.

Data over a 4-year period indicated that yield increases

were directly proportional to the size of sets used for planting. Extra large sets, however, tended to produce hollow
roots. In a 7-year study on planting dates, the earliest
plantings produced the highest yield, but planting could be
delayed until May 1 without a serious reduction in yield.
Enlargement of the primary root was found to occur mainly in
the cooler temperatures of September and October. Data from
fertilizer tests have been inconclusive, but mineral composition studies (5) indicate that plant analysis to monitor a
soil fertility program should be limited to sampling of the
leaf blade at the fern stage of growth. Application of plant
growth regulators to sets for retarding or preventing secondary root growth near the crown or to stimulate root growth
at the basal end have not been sucessful. Cultivars from
different sources have been tested almost yearly since 1937.

III. BREEDING

     Horseradish production in the U.S. is confined mainly to
3 cultivars, Common, Swiss, and Big Top Western (2). This
narrow genetic base is especially vulnerable to heavy crop
loss from diseases, insects, or other causes. For this reason, a breeding program was initiated with the objective of
expanding the genetic base by developing additional cultivars
(6, 7). Since farmers must grow and maintain their propa-
gating stock, it is theoretically possible that each farmer
could grow a different cultivar given sufficient choice.
     To initiate the program, a gene bank was established from
plants collected for us by the U.S. Department of Agriculture
from several areas in Europe. Additional germ plasm came
from seedlings originally bred at the University of Wisconsin
AES (8) and from other sources in the United States. The
gene bank, including new seedlings from the current breeding
program, now contains about 200 genotypes.
     To assist in the choice of germ plasm for perservation
and breeding, 2 taxonomic studies were made (6, 7, 9). The
results indicated that cultivars could be grouped into 3
types using 2 highly diagnostic characters, basal angle of
leaf blade and degree of leaf crinkling. These 2 characters
were found to be associated with 2 diseases, turnip mosaic
and white rust. Plants of type I have smooth leaves that
taper acutely at the base, are resistant to the white rust
disease, and show no symptoms of turnip mosaic. The cultivar
Big Top Western belongs to this group. Type II has smooth
to slightly crinkled leaves that are rounded at the base, and
turnip mosaic is expressed as a ringspot symptom. 'Swiss' is
a member of this group. Type III has crinkled, cordate

leaves and is susceptible to turnip mosaic, expressed as mottling, and to white rust. 'Common' belongs to this group.
It is of interest that the leaf shapes of Types I and III are
illustrated in the herbals of the 16th Century (10, 11).

The operation of this program differs somewhat from most
crop breeding programs. Cultivars from a typical breeding
program are made available only after extensive testing and
formal release. In our program selection and introduction of
new cultivar is a joint effort between the University of
Illinois and Illinois horseradish growers. The growers provide time, labor, and land, and also funds. Close cooperation is possible because only about 40 Illinois farmers located in a relatively small area produce horseradish.

A typical cycle for breeding, testing, and introduction
of new cultivar takes 8 years. Twenty to 30 genotypes are
selected the first year from the field test plots for seed
production. Crowns of primary roots are stored at $2^{o}C$ and
the following April, they are planted in the greenhouse.
Flowers appear in 3 weeks and by mid-June seeds are ripe
enough for harvest and immediate planting. We make some
controlled pollinations, but most seeds are produced through
insect pollinations, usually flies. Seeds are also obtained
from our perennial gene bank. Ninety-five percent of genotypes tested set viable seeds, with sterility being found
mainly in the crinkled-leaf type.

The new seedlings are grown in the greenhouse for 2
months, then transplanted to the field. Seedlings are dug
the following year and plants that have produced healthy
roots are saved for the first cooperative grower trial. One
set is selected from an individual plant for either planting
at Urbana, or in a growers field. The following fall research
and extension personnel, together with 15 to 20 growers,
harvest, clean, trim, and weigh each entry. The growers
judge and rank each entry for quality. From the results of
their combined judgment, 3 to 5 sets per cultivar of the top
selections are saved for further testing. Advanced selections
are tested for 3 more years in 10 root plots on grower farms.
A cultivar that survives this test is increased and each
grower receives 10 to 30 roots for testing on his particular
farm. New cultivars are also made available on request to
out-of-state growers.

For the first few years of the breeding program, about
150 seedlings were produced annually. Only 2 cultivars
merited release, but both had undesirable traits. One cultivar, Illinois 830a, lacked the desired pungency and the
other, Illinois 473a, was difficult to clean due to the
abundance of fine root hairs on the marketable root. For this
reason, Illinois 473a was nicknamed 'Old Hairy.' The program

was expanded in 1973, and, since then, we have tested about 1200 new seedlings each year.

In 1975 a disease called brittle root destroyed many seedlings as well as commercially grown cultivars at both Urbana and East St. Louis.

To date we have not directed our breeding program towards resistance to a particular disease. The reasons for this non-specific disease selection will be explained in the section on diseases.

IV. DISEASES

The nature and control of diseases of Illinois horse-radish have been studies by members of the Department of Plant Pathology since 1933. Three of these diseases, white rust, turnip mosaic, and brittle root are of major economic impor-tance.

The most damaging foliage disease is white rust, caused by the fungus *Albugo candida* (Pers. ex. Chev.) Kuntze. Ex-tensive leaf damage due to the disease prevents normal root growth and results in reduced yields. Since the disease spreads from the crown of the primary root into secondary roots, Endo and Linn (12) recommended that only healthy sets taken from the basal end of the primary root be saved for planting the next crop. Also, all systematically infected plants should be removed from the field and destroyed in the spring. Fungicides are recommended during the growing season. By following these procedures, the growers have been able to control the disease.

Certain genotypes of horseradish are known to be resis-tant to white rust (13, 14), but to date we have not bred specifically for resistance to the disease.

Turnip mosaic virus occurs world wide in an extended host range (15). It can be transmitted from horseradish to other plant species by aphids or by mechanical means (16). In Illinois, Chenulu and Thornberry (17-19) have successful-ly transmitted the disease from horseradish to at least 16 other widely diverse plant species. Foliar symptoms on horseradish include mosaic mottling and chlorotic rings of the leaf blade, and black streaking of the petiole (20).

Holmes (21) attempted to eliminate the virus in horse-radish plants by chemotherapy applications and he concluded that all widely grown cultivars were infected. The disease is apparently not seed transmitted. Seedlings raised in the greenhouse show no symptoms, but after these plants have been transplanted to the field, they usually become infected with-in 2 months.

Recently, G. M. Milbrath, a virologist in Plant Pathology, tested several cultivars of horseradish and found one
plant of 'Big Top Western' that was negative for the presence
of turnip mosaic virus. If this plant is virus free, then
resistance to the disease probably exists in horseradish
germplasm, because it is difficult to explain otherwise how
a plant could remain virus free after years of vegetative
propagation in commercial fields.

Prior to 1951, the East St. Louis growers purchase new
sets every 3 or 4 years from the Chicago area or Wisconsin,
because they felt that their homegrown stock gradually de-
teriorated in quality. To solve this problem, Thornberry and
Linn demonstrated that homegrown sets selected from large
healthy plants rather than from sets selected at random would
continue to produce good crops. By following careful selection procedures, in addition to better disease and insect
control practices, the growers now propagate sets continually,
rather than buy new sets periodically.

Brittle root, whose etiological agent has yet to be determined conclusively, is probably the most destructive disease of horseradish (22). Infected plants develop a general
chlorotic condition of the leaves and a collapse of the aerial portion of the plant. The disease gets its name from
the symptoms that occur in the roots. The phloem is usually
a dark brown and a dark ring forms around the vascular cy-
linder. Starch accumulates in the roots which become turgid
and snap when bent, hence the term brittle root.

The disease occurs sporadically and it caused extensive
crop losses in Illinois in 1935-36 (23), in 1953-54 (24, 25)
and in 1975. Thornberry & Takeshita (24) cite circumstantial evidence, but no proof, that the beet leafhopper,
Circulifer tenellus Baker, transmits the sugar beet curly-top
virus, *Ruga verrucosans* Carsner & Bennett, to horseradish
resulting in the brittle root symptom. This insect as well
as several other leafhoppers were found in East St. Louis
horseradish fields in 1936 (26) and in 1953-54 (27).

Environmental factors necessary for the rapid spread of
the disease are unknown. Plants infected late in the season
may show no disease symptoms, but sets from them planted the
following spring (28) will develop brittle root. In most
years a few brittle root plants can be found in cultivated
fields of horseradish, but the disease does not occur in
epidemic proportions.

Whether resistance to brittle root exists in horseradish
is unknown, but breeding for resistance will be virtually
impossible until we know its cause and mode of transmission.
In 1975, the disease did not systematically destroy plants
within confined areas, but it appeared to spread more or less

at random. It was not possible, then, to determine whether
a seedling, represented by a single plant, was resistant or
whether it had escaped infection.

Because of the destructive nature of the disease, the
Department of Plant Pathology has initiated a research project
to study brittle root and the viruses that occur in horseradish. The work is being supported in part with financial aid
from the horseradish industry, including growers, brokers,
and processors.

V. INSECTS

Research on insects associated with horseradish in Illinois is conducted by Agricultural Entomology and the State
Natural History Survey.

Many species of insects attack horseradish, but only a
few cause economic damage. In studies from 1947 to 1954,
Petty (24) found 42 species in horseradish fields. Six species decreased yield of horseradish. These species included
onion thrip, *Thrips tabaci* Line; beet leafhopper, *Circulifer
tenellus* Baker; green peach aphid, *Myzus persicae* Sulz.;
mealy plum aphid, *Hyaleptorus arundinis* F.; diamondback moth,
Plutella maculipennis Curt.; and southern cabbage worm,
Pieris protodice B. & L.

Petty (27) also observed 2 species of flea beetles,
horseradish flea beetle, *Phylletreta armoracies* Koch and
crucifer flea beetle, *P. cruciferae* Csc., in sufficient numbers to cause damage to horseradish. Previously, Compton (29)
stated that the horseradish flea beetle was the only insect
pest of horseradish that required control measures, and Davis
(30) reported injury to horseradish from the insect in 1909.

In some years, two spotted mites are a problem. No
miticides are labeled for use on horseradish, but malathion
applications will reduce mite populations. In general, mite
populations increase after use of carbaryl to control caterpillars. This treatment reduces mite predators.

VI. PROCESSING

Processors manufacture fresh horseradish continually
from supplies of root held in cold storage. High quality
horseradish sauce is white and pungent. In storage the sauce
gradually loses its pungency, becomes discolored, and develops an earthy odor and taste. Consumers may refrain from
purchases because they know from experience that the product

may deteriorate before it can become completely consumed.

Most horseradish sauce contains comminuted root, acetic acid from vinegar, and salt. Fresh whipped dairy cream may be added to lengthen shelf-life.

The chief pungent principle of horseradish is allyl iso-thiocyanate, allyl-NCS (31). In a raw root, the mustard oil is bound with glucose and sulfate ion as a water soluble glu-coside, Sinigrin. When raw root is macerated, the enzyme myrosinase and the substrate sinigrin are united and allyl-NCS is developed instantly (32, 33).

The Department of Food Science has been active in re-search concerning processing and preserving horseradish. Weber (34, 35) developed a method of steam blanching and freeze dehydration for preserving horseradish powder. Spata (36) found that the pungency of rehydrated horseradish powder could be increased by addition of freshly dried root material or L-ascorbic acid.

Weber et al. (32) found that frozen storage at $-18^{o}C$ was an effective means of retarding quality degradation of pro-cessed horseradish; and at $7^{o}C$, food additives, such as pro-pylene glycol alginate, added to cream style horseradish provided an effective retardation of pungency loss. In another study, they (33) found that prepared horseradish with 98 percent blanched and 2 percent raw root had higher pungency and better color retention than 100 percent fresh root after 8 weeks storage. They explained that blanching inactivates the enzyme system and with the addition of a small amount of fresh root containing the enzyme, the sini-grin of the blanched product is converted to allyl-NCS.

Aung (37) found that the substrate of tartaric acid for vinegar improved pungency retention and the addition of L-ascorbic acid in the formulation improved color retention.

Many processors when grinding horseradish roots dis-charge the ground material directly into acid solution (vinegar) or pickle. The acid essentially arrests the en-zyme action which converts sinigrin to the pungent principle. Studies by the Department of Food Science have found that grinding roots with crushed ice prior to addition of vinegar produces a highly pungent horseradish sauce. Grinding a mix-ture of roots and crushed ice materially reduces pungency loss by substantially lowering the temperature of the ground mixture which in turn reduces the volatility of the pungent principle.

VII. STORAGE

Horseradish roots are stored for year-round processing in refrigerated storages without serious deterioration in quality. Temperatures of 0° to 2°C and relative humidity of 90 percent are recommended. Plastic bag liners have been found to reduce water loss. University of Illinois studies showed that bag and root temperatures ranged from 1.5 to 2.0°C higher than air temperatures, and that cooling was adequate with proper stacking.

Roots are cleaned of excess dirt and stored without washing or chemical treatment. Tests showed that unwashed roots had no more storage rots than those washed or chemically treated.

Similar results were obtained from treatments on sets stored for the next crop. Good sanitary practices in handling the sets, and storage in either refrigerated coolers or buried in soil, has been satisfactory. In our breeding program, we store roots in plastic bags at 2°C, and under these conditions, we have kept root material in storage for 2 years without loss of viability.

VIII. FUTURE

The average per capita consumption of horseradish in the United States is less than 30 g annually. If the industry is to prosper, the consumer must be given the product in a form that does not deteriorate rapidly and must be educated on how to use and store it. Only then will consumers experience the joy that a fresh high quality product can add to good eating (38). For example, some people add a bit of fresh horseradish sauce to enhance the flavor of pizza. The promotion of this little known "secret" might result in a considerable increase in the demand for horseradish.

Horseradish growers tell us that they can increase production, but the market is limited. Research is needed in areas such as promotion and advertising, open-dating, new products, and mini-packaging to increase frequency of use. Test sales in individually dated containers in amounts suitable for individual meals might lead to increased consumption.

ACKNOWLEDGEMENTS. I thank the many University of Illinois
staff members and the Illinois horseradish growers who have
contributed the information and help necessary for writing
the manuscript. A special thanks is due J. W. Courter, C. C.
Doll, H. J. Hopen, M. B. Linn, G. M. Milbrath, H. E. Nelson
and R. Randell for their suggestions during the preparation
of the manuscript.

IX. REFERENCES

1.   De Candolle, A., "Origin of cultivated plants." Hafner
     Publishing Co., New York, 1959. (Reprint of Sec. ed.,
     1886).
2.   Courter, J. W., and Rhodes, A. M., Econ. Bot. 23, 156-
     164 (1969).
3.   U.S. Tariff Commission. "Crude horseradish" Rept. on
     Excape-Clause Invest. No. 7-88, provisions of sect. 7,
     Trade Agreements Ext. Act. of 1951 as amended. Wash-
     ington, D.C., 1960.
4.   Hopen, H. J., and Doll, C. C., North Central Weed Con-
     trol Con. (Abstr.) 24, 54. (1969).
5.   Peck, T. R., Courter, J. W., Rhodes, A. M., and Walker,
     W. M., Agron. J. 61, 526-527. (1969).
6.   Rhodes, A. M., Courter, J. W., and Shurtleff, M. C.
     Ill. State Acad. Sci. 58, 115-122. (1965).
7.   Rhodes, A. M., Courter, J. W., Shurtleff, M. C., and
     Vandemark, J. S., Ill. Research University of Illinois
     AES, Fall. 7(4) 17. (1965).
8.   Weber, W. W., J. Hered. 40, 223-227. (1949).
9.   Rhodes, A. M., Carmer, S. G., and Courter, J. W., J.
     Amer. Soc. Hort. Sci. 94, 98-102. (1969).
10. Fuchs, Leonhart. "New Kreuterbuch" Michael Isingrin.
     Basel. 1543.
11. Gerard, J., "The herball or generall historie of plants"
     John Norton, London, 1597.
12. Endo, R. M. and Linn, M. B., Ill. Agr. Exp. Sta. Bull.
     655:1-56. (1960).
13. Hougas, R. W., Rieman, G. W., and Stokes, G. W.,
     Phytopathology 42, 109. (1952).
14. Shurtleff, M. C., Scott, D. H., Linn, M. B., and Rhodes,
     A. M. "White rust of horseradish." Report on plant
     diseases No. 960. Dept. of Plant Pathology, Coopera-
     tive Extension Service, Univ. of Illinois, and U.S.
     Dept. of Agr., 1967.

15. Tomlinson, J. A., "Turnip mosaic virus" C.M.I.A.A.B. Description of plant viruses No. 8. Commonwealth Agric Bur. and Assoc. of Appl. Biol. W. Culross and Sons, Ltd., Perthshire, Scotland, 1970.

16. Hoggan, I. A., and Johnson, J., *Phytopathology* 25, 640 (1935).

17. Chenulu, V. V. "Studies on host range, bioassay and properties of turnip mosaic virus (*Marmor bassicae* H.) from horseradish" Ph.D. Thesis. Dept. of Plant Pathology, Univ. of Illinois, Urbana, 1959.

18. Chenulu, V. V. and Thornberry, H. H., *Current Sci.* 31, 516. (1962).

19. Chenulu, V. V. and Thornberry, H. H., *Plant Dis. Reptr.* 48, 259 (1964).

20. Pound, G. S., *J. Agric. Res.* 77, 97 (1948).

21. Holmes, F. O., *Phytopathology* 55, 530. (1965).

22. Thornberry, H. H., *Ill. Research* Univ. of Illinois AES, Spring. 3(2), 14 (1961).

23. Kadow, K. J. and Anderson, H. H., *Plant Dis. Reptr.* 20, 288 (1936).

24. Thornberry, H. H. and Takeshita, R. M. (Endo), *Plant Dis. Reptr.* 38, 3 (1954).

25. Thornberry, H. H., *Plant Dis. Reptr.* 39, 801 (1955).

26. Delong, D. M. and Kadow, K. J., *Jour. Econ. Ent.* 30, 210 (1937).

27. Petty, H. P. "The insect pests of horse-radish in southern Illinois." Ph.D. Thesis. University of Illinois, Urbana, 1955.

28. Kadow, K. J. and Anderson, H. H., *Ill. Agr. Exp. Sta. Bull.* 469, 529 (1940).

29. Compton, C. C., *Ill. Agr. Exp. Sta. Circ.* 295, 1 (1925).

30. Davis, J. J., *J. Econ. Ent.* 3, 180 (1910).

31. Stoll, A. and Seebeck, E., *Helv. Chim. Acta.* 31, 1432 (1948).

32. Weber, F. E., Nelson, A. I., Steinberg, M. P., and Wei, L. S., *Food Technol.* 23(9), 103. (1969).

33. Weber, F. E., Nelson, A. I., Steinberg, M. P., and Wei, L. S., *Food Technol.* 23(9), 107. (1969).

34. Weber, F. E. "Pungency and physical properties of air and freeze dehydrated horseradish" M.S. Thesis. Dept. of Food Science, University of Illinois, Urbana, 1963.

35. Weber, F. E. "Physical and chemical characteristics of bottled fresh and dehydrated horseradish" Ph.D. Thesis. Dept. of Food Science, University of Illinois, Urbana, 1964.

36. Spata, J. M. "Process and storage variables affecting quality of freeze-dried horseradish" M.S. Thesis. Dept. of Food Science, University of Illinois, Urbana,

1965.

37. Aung, Thein. "Improved pungency and color stabilization
     of horseradish sauce by chemical additives and pro-
     cessing treatments" Ph.D. Thesis. Dept. of Food Sci-
     ence, Unov. of Illinois, Urbana, 1967.

38. Doll, C. C., Courter, J. W., Acker, G., and Vandemark,
     J. S. "Illinois horseradish, a natural condiment"
     College of Agr., Univ. of Illinois, *Coop. Ext. Serv.
     Circ.* 1084, 1973.

DIOSCOREA - THE PILL CROP

Norman Applezweig

Norman Applezweig Associates
New York, New York

I.   INTRODUCTION

The use of the word "pill" in the title refers to the oral contraceptives which were created by steroid chemists and physiologists who determined and then modified the structures of the sex hormones which naturally occur in mammals. The word "pill" is also intended to cover the natural and modified steroid hormones of the adrenal cortex, namely the cortico-steroids. Together these categories of drugs represent a re-tail market of almost a billion dollars and are among the most important medical weapons yet devised.

Our ability to produce these drugs in quantities and at costs which render them available can be traced directly to the exploitation of the steroidal sapogenins contained in dioscorea and this, in turn, was made possible by the work of one man, Russell E. Marker. One of the most gifted organic chemists of our time, Marker was also an amateur botanist and obsessive searcher for plant sources of steroid hormones (1).

The term steroids covers the entire area of compounds de-rived from sterols. The original and most important sterol is cholesterol which is the source of human atherosclerotic disease but also the building block in the adrenal glands, the male and female sex gonads, for the corticosteroids or stress hormones and the male and female sex hormones. Cholesterol also serves as the insulating material for the entire nervous system.

Plants also have a sterol economy based upon phytosterols, close relatives to cholesterol, which are found in the oil seeds and in some plants as alkaloids, amines and steroidal sapogenins. With the high cost of cholesterol obtainable from animal sources and the poor yields involved in its chemical conversion to steroid hormones in the 1930s, plant sterols seemed to be logical raw materials for hormone production.

149

A.   Marker's Development of the Diosgenin Potential

In 1935, when Marker embarked on his search, he was working on a fellowship supported by Parke, Davis & Co. at Pennsylvania State College. His stipend was a remarkable $1,800
per year! Parke, Davis sent Marker steroid fractions accumulated from the extraction of animal pregnancy urines and also
a kilogram of crude cholesterol. Marker learned to work with
steroids in these early experiments which finally culminated
in more than fifty papers published between 1935 and 1938.

His first work on sapogenins was published in 1939 (2).
In this paper Marker proposed a new structure of the side
chain of sarsasapogenin. He converted sarsasapogenin to
Δ_{16}-pregnenolone acetate by what is now known as the Marker
degradation which removes six carbons down to. but no further
than the pregnane (C_{21}) nucleus. This is essentially the procedure still used by the entire steroid industry today. Also,
he showed that his pregnane derivatives from sapogenins were
identical with those obtainable from pregnancy urine and con-
firmed that plant sapogenins could easily produce progesterone
and other mammalian sex hormones.

With this evidence Marker undertook an arduous search for
plant sources of sapogenins which lead him through various
parts of the United States and finally to Mexico. There he
came upon specimens of dioscorea in the state of Vera Cruz.
His total work involved an investigation of over 40,000 kilos
of plants comprising more than four hundred species and the
isolation and identification of twelve new sapogenins (3). As
a result of this work he became convinced that diosgenin from
Mexican dioscorea represented the optimum raw material for
hormone manufacture. Unfortunately, he was not able to interest Parke, Davis & Co. in engaging in such a business since
they were wary of trying to do anything in Mexico. Therefore,
Marker set out on his own to start the manufacture of progesterone in Mexico from diosgenin obtained through the collection of wild dioscorea in the Mexican jungles. In the
course of setting up this business he joined in a partnership
to form Syntex and then later went on to start other companies.

The Mexican steroid industry and, for that matter, the
world hormone industry is a living monument to the living
Russell Marker who, despite rumors to the contrary, is alive
and well and living in State College, Pennsylvania. The
rumors undoubtedly came about because when he felt he had ac-
complished what he had set out to do after five years of production and research in Mexico, he retired in 1952 and turned
his interests elsewhere. Marker never earned very much for
his activity. The industries he started are worth a great
deal of money but he feels he was amply rewarded by having
found the sources for the production of steroid hormones in

quantity, at low prices, developed the process for manufacture
and put them into production.

The ideal raw material for the production of contraceptive steroids was and is diosgenin. If Marker had not unearthed Mexican diosgenin as the most versatile and available
steroid raw material it is probable that we would not have had
oral contraception in our time, or even in our century. Of
all the steroid materials available at the time of Marker's
explorations, only the sapogenins presented the possibility of
easy conversion in high yields to both androstanes and preg-
nanes. Thus testosterone, progesterone and the estrogens
became available at low cost from a single compound relatively
easy to extract from abundantly available natural sources.

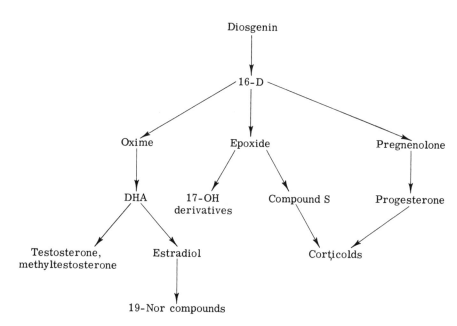

*Fig. 1. "From Steroid Drugs by Applezweig. Copyright
1962, McGraw-Hill. Used with permission of McGraw-Hill Book
Company."*

The Mexican steroid industry, based upon the diosgenin
potential, grew suddenly into a giant when Upjohn developed
their fermentative oxidation of progesterone, enabling diosgenin intermediates to be used for the production of corticosteroids (4). It was this single fact that turned what was a
fine chemical operation into a major industrial undertaking
and, in turn, made it possible for diosgenin intermediates to
be produced in such quantity and such low cost that oral con-

traceptives became a possibility. The industry developed
superb technology, excellent technicians and brought increasing investment and income to Mexico. In addition to the low
cost, the industry's research on corticosteroids led to the
development of modified progestagens which became oral contraceptives.

B.   Diosgenin Versus Alternate Raw Materials and Routes

1.   *Alternate sources of steroid raw materials*

     Other steroid raw materials were available when Mexican
diosgenin intermediates became articles of commerce. Cholesterol and stigmasterol were already used commercially. Hecogenin, tigogenin, sarsasapogenin, smilagenin, solasodine,
tomatodine and, of course, bile acids were all explored as
potential competitors for diosgenin. But the early start and
superb performance of the Mexican steroid industry made most
of these alternative seem economically risky and, until the
cost of diosgenin began to rise in recent years, they did not
play a competitive role in the world steroid picture.

     There were, of course, a few exceptions. Upjohn, in
order to protect its prime position in corticosteroids, made
a relatively huge investment in the separation of stigmasterol
from soy sterols (5). It is certain, however, that their stigmasterol operation helped to drive down Mexican prices and
Upjohn, in turn,helped by buying a good portion of Mexico's
production. Upjohn retained its leading position in corticosteroids but the fermentation processes were licensed to others,
some of whom improved on the original process and almost all
of the new producers turned to Mexican intermediates as their
raw material.

     Hecogenin from sisal was also a minor exception. Glaxo
(6) in England, using some processes developed by Syntex (7),
turned to African hecogenin for their corticosteroid raw material since it was to their advantage to deal in Sterling areas
despite higher costs of production.

     Bile acids were the first raw materials used for corti-
sone production (1b,c). Their limited availability soon made
them unattractive, especially after the Upjohn fermentation
discovery. However, when the great demand for bile acids decreased, certain companies, such as Roussel and Organon, were
able to buy their needs carefully so that cortisone production
from bile acids continued and, indeed, today accounts for a
healthy portion of the world's corticosteroids.

     In Eastern Europe, solanum was, at first, the raw mater-
ial of choice but because of the efficiency of the Mexican
diosgenin production this raw material was abandoned and Mexican intermediates were instead used in the Hungarian corti-

Steroid Product Development

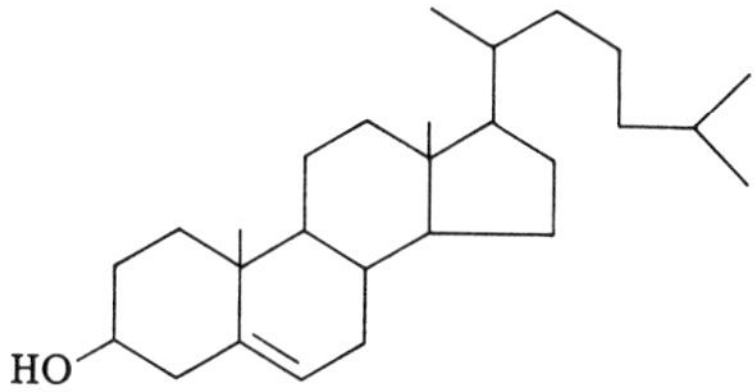

Cholesterol

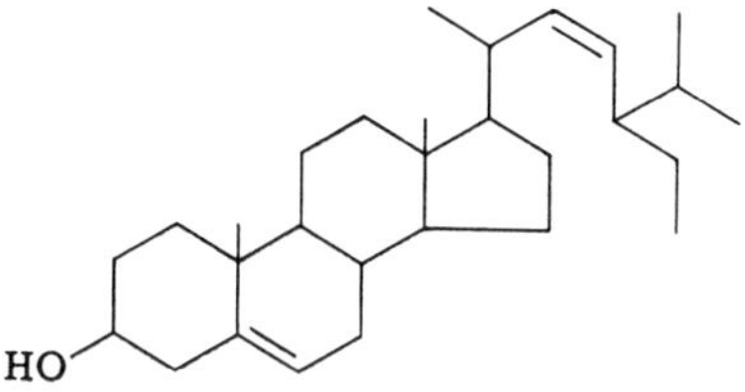

Stigmasterol

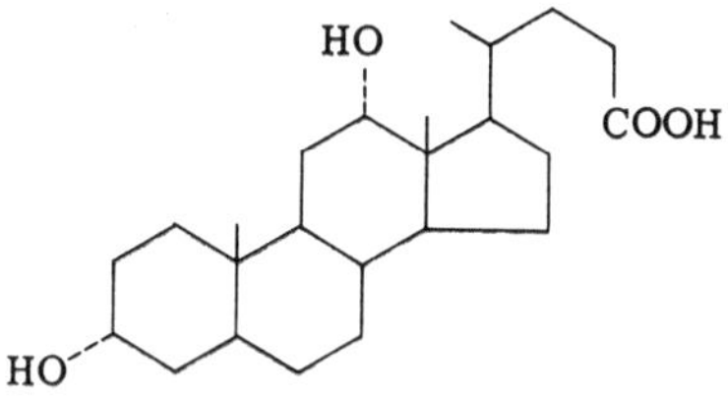

Desoxycholic acid

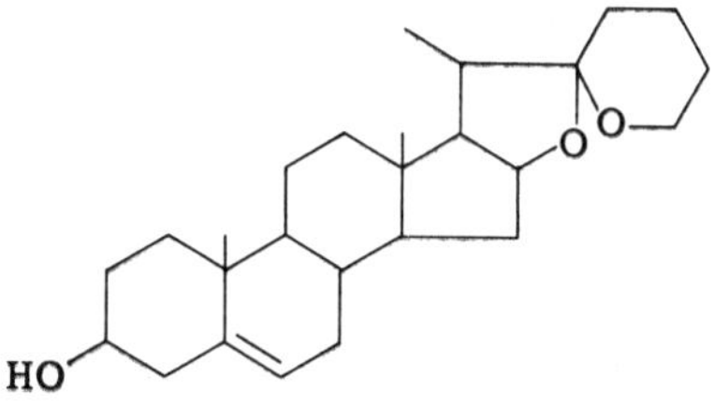

Diosgenin

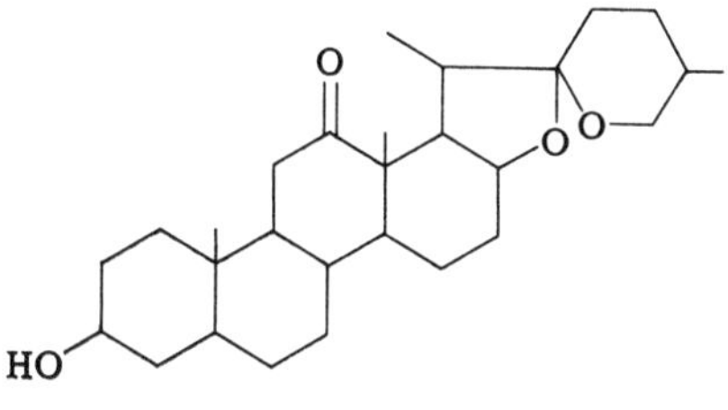

Hecogenin

Fig. 2. "Reprinted from the May 17, 1969 issue of
Chemical Week by special permission. Copyright 1969 by
McGraw-Hill, Inc."

costeroid fermentation processes. It is evident that diosgenin
attained its prime position because it was the first with the
most. It was produced in greater quantity than any other raw
material and, therefore, was able to outprice any potential
competitor.

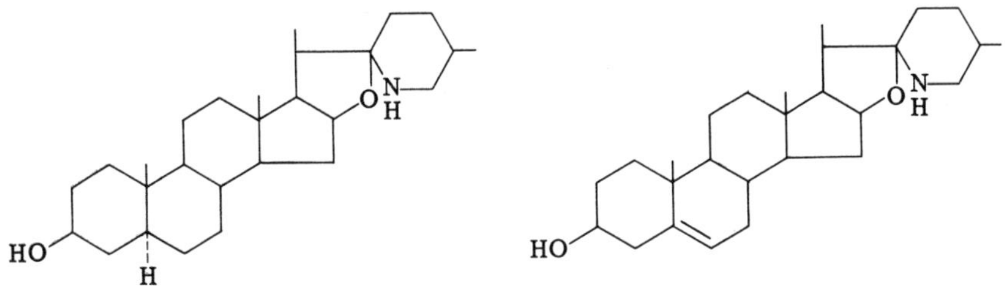

Tomatidine equivalent to neotigogenin   Solasodine equivalent to diosgenin
Soladulcidine equivalent to tigogenin   Solauricidine equivalent to yamogenin

Fig. 3. "Reprinted from the May 17, 1969 issue of Chemical Week by special permission. Copyright 1969 by McGraw-Hill, Inc."

TABLE 1

World Steroid Raw Materials Sources (8)

Country	Material	Production* '63	'68
Mexico	Diosgenin	375	500
	Smilagenin	–	10
United States	Stigmasterol	60	150
Guatemala	Diosgenin	10	30
Puerto Rico	Diosgenin	–	20
France	Total synthesis	–	50
	Bile acids	20	50
Germany and Netherlands	Cholesterol and bile acids	5	10
Africa	Hecogenin	20	40
India	Diosgenin	10	30
China	Diosgenin	–	80

*Production figures expressed as tons of diosgenin.

2. *New developments in utilization of contraceptive steroids*

When contraceptive steroids became a reality, the Mexican steroid industry again became the supplier of choice because of its prime position. Total synthesis has been developed in various laboratories but the early sales of contraceptive steroids were relatively small and unless total synthesis can be scaled up to very large volumes, it is difficult to create a basic industry which could be competitive with the low cost naturally derived materials.

In the last fifteen years there have been many exciting events in the contraceptive steroid field: (1) The number of users grew at a fantastic rate, beyond any of the early predictions, so that from kilos of progestagens we started to think in terms of hundreds of kilos, tons, and then in tens of tons. (2) Simultaneously, however, daily doses were reduced from 10 mgs. to 5, to 2.5, to 2, to 1, to 0.5 and, finally, to 0.35. Since these these events occurred at different times in different places in the world, it would take a computer to calculate exactly where the curves for users and usage coincided. It is now clear, however, that the tonnage requirements for progestagens are considerable, although they in no way compare with the usage of corticosteroids. Nevertheless, it may now seem attractive to consider other raw material sources, total synthesis on a large scale, and indigenous production of contraceptive steroids and their raw materials by various areas in the world wishing to effect either cost, availability or national independence. (3) All that remains to convert these possibilities into realities are: a) an insufficient supply of diosgenin intermediates; b) an increase in cost of such materials; and, c) uncertainty as to either cost or supply. I am concerned that this has happened in the case of the Mexican steroid industry. A vacuum was created and it is being filled by other means.

3.   *Recent changes in the relative costs of raw materials*

In January 1975, the Mexican government took over the collection of dioscorea and set new prices and allocations for the industry. Producers of Mexican steroid intermediates were faced with a sudden four-time increase in raw material cost and an uncertainty about amounts that would be available. The possibility of a steroid shortage became imminent. Companies throughout the world who had based their processes on Mexican intermediates had to curtail their production because they could no longer compete with others using alternate raw materials. The demand for product from the producers using stigmasterol, hecogenin, bile acids and total synthesis outdistanced their ability to increase their production facilities. An intense inflation appeared to be in the making and the incentive to fill the vacuum became greater.

a.   *Other sources of diosgenin*. Mexico is not the only source of diosgenin. Guatemala, Costa Rica, India and China have also been factors, though minor thus far, in diosgenin supply. China is a major producer and while it does offer some diosgenin in the world market, it uses most of its raw material for its own industry and its own people. Information indicates that supplies of uncultivated dioscorea in India are

diminishing and they are looking either to cultivation or to
other raw material sources. Cultivation of dioscorea is possible but costs are high. Diosgenin can also be made from the
Fenugrek plant which is cultivated as a flavoring agent in
many parts of Africa, and also from a wild plant, *Balanites*,
which can also be cultivated as a food crop in arid areas (9).
        b. *New sources of hecogenin* New and relatively large
sources of hecogenin are now on stream from new areas of the
world. Most promising is the potential for large scale production in Haiti. This will certainly contribute to a normal-
ization of corticosteroid cost and supply (10).
        c. *Microbiological production of androstanes* The micro-
biological processes for degradation of cholesterol and sitosterol to androstanes leading to androgens, estrogens, progestagens and spironolactone were dug out of the patent files,
taken off the drawing boards and put into industrial production (Table 2 & 3). Mitsubishi Chemical Industries (MCI),
with almost unlimited quantities of cholesterol from the wool
grease of the Japan wool processing industry (10). Schering,
A. G. and Searle, using sitosterol from soy sterols, are all
building industries competitive to Mexican diosgenin (Table 2
& 3).

d. *Total synthesis* Total synthesis of norethindrone by
Roussel of France is now a large scale reality since both the
market and the prices have grown to where the original Velluz
process is now fully competitive and capable of economic expansion without great risk (10). Also active in total synthesis now are laboratories in Germany, both East and West,
Hungary, Switzerland, China and the U.S. However, since total
synthesis of corticosteroids is not feasible with existing
methods, natural materials remain the major source of world
steroid production.

e. *Solanum steroids* Solasodine is now an attractive competitor for diosgenin since it possesses the same versatility
and can be grown rapidly in areas of cheap labor. The growth
of solanum plants is certain to proliferate in areas wishing
to have indigenous raw material supplies (9). Ecuador, Brazil,
New Zealand, Australia, Indonesia, Hungary and Israel are examining their own prospects for solanum alkaloid production.
Other steroidal sapogenins are also being re-examined but it
may already be too late for new raw material sources to be
started at this time since the vacuum is being rapidly filled
and the threat of a steroid shortage may soon turn into a
steroid glut.

TABLE 2

Microbiological Conversion of Cholesterol and Phytosterols (11)

I.   *Mitsubishi Chemical Industries process* (Arima, et al). U.S.
     Patent 3,388,042 (1968). Priority Japanese application
     1964. Cholesterol, sitosterol and stigmasterol are con-
     verted to androstenedione (AD) and androstadienedione
     (ADD) by action of a micro-organism* having a sterol
     decomposing ability in the presence of a compound capable
     of forming a chelate with iron and/or copper.
     (* wide variety of micro-organisms covered)

II. *Gist-Brocades, Delft process I* (van der Waard, et al).
     British Patent 1,113,887 (1970). Similar to above but
     fermentation carried out in presence of nickel, cadmium,
     cobalt and/or selenite ions by micro-organisms of the
     genus *Mycobacterium*.

III. *Gist-Brocades, Delft process II* (van der Waard, et al).
     U.S. Patent 3,487,907 (1970). Same as above except using
     micro-organisms other than genus *Mycobacterium*.

IV. *Hungarian process* (Wix, et al). Hungarian Patent 153,831
     (1965). Used 8-hydroxyquinoline as a chelating agent to
     inhibit cleavage of the steroid nucleus via 9-hydroxyla-
     tion during conversion of cholesterol ergosterol or
     stigmasterol to ADD. Use Mycobacteria varieties.

V.   *Searle process I* (Krachy, et al). U.S. Patent 3,684,657
     (1972). Preparation of AD and ADD from 17-alkyl steroids
     (cholesterol and phytosterols) using *Mycobacterium* sp.N.
     R.R.L.B.-3683 or enzymes thereof.

VI. *Searle process II* (Marsheck, et al). U.S. Patent
     3,759,791. Selective preparation of AD from 17-alkyl
     steroids using *Mycobacterium* sp.N.R.R.L.B.-3805 or
     enzymes thereof.

TABLE 3

Comparative Costs Per Kilo (11)

(Based upon published prices & yields)*

Raw Materials/K		Contraceptive Steroids (Equiv. Norethindrone)	Corticosteroids (Equiv. Hydrocortisone)
Diosgenin	$ 15	$ 150	$ 90
	60	600	360
	120	1,200	720
Hecogenin	120	---	720[a]
Stigmasterol	45	---	315
Stigmasterol	(sitosterol recovered)		203
Sitosterol	4	84	---
Cholesterol	16	42	---
Beef Bile	10	---	50[b]

*Solvents, reagents, nutrients, power, labor and overhead vary
with each process and are usually many times the cost of the
raw material input.
[a]Useful mainly in production of dexamethasone and
betamethasone.
[b]Bile process has highest labor and reagent cost.

C. *Present and Future Raw Material Needs*

What are the world's present requirements and what are
the potentials for fulfilling them in the coming years? For
contraceptive steroids, present use is relatively small because of modern low dosage preparations. I estimate that we
now use between 60 to 180 tons of diosgenin equivalent; by
1985, it could rise to between 200 and 500 tons; and, hopefully, within the following ten years may be as much as 3,000
tons (Table 4). While it is true that at present any one of
the companies in Mexico might supply enough diosgenin intermediates to take care of the entire world contraceptive
steroid demand, it must be stressed that these companies cannot live on contraception alone. Most of their products are
used for corticosteroids and it would be neither prudent nor
practical to concentrate on one area and lose out in the other.
Contraceptives cannot be made to be the tail that wags the dog
in the steroid industry. The consumption of steroids for contraception represents only about 5% of the steroids produced
(Table 5).

TABLE 4

Contraceptive Steroids (11)

	World excl. China		China	
# of women (millions)	30 —	40	20 —	25
@ 1 mg.	8,190 K	10,920 K	5,460 K	6,825 K
@ 0.5 mg.	4,095 K	5,460 K	2,730 K	3,412 K

	Total World Needs -- Present	
	Progestagens	Diosgenin
min.	6,825 K	69 tons
max.	17,745 K	180 tons

Total World Needs of Diosgenin -- Future	
	By 1985 (3x expansion)
min.	201 tons
max.	540 tons
	By 1995
Approx.	2,700 tons
	(200 million women)

Corticosteroids presently consume 1,200 tons of diosgenin equivalent, and will grow by at least 10% per year in the next ten years. If a shortage occurs, the corticosteroids will be mainly affected.

With a return to normalcy in steroid costs which will follow the introduction of new sources of supply, the agricultural and veterinary uses of steroids will take on a new meaning both in meat production and in animal health which may well rival some of the present uses of steroids in the human field.

D.   *Will There Be Enough Steroids?*

Already we see that world needs for steroids far exceed the potential diosgenin production of Mexico. In 1974, Mexico produced approximately 600 tons of diosgenin. A remaining

TABLE 5

Corticosteroids and Spironolactone (11)

Present	150 tons	Diosgenin Equiv.	1,200 tons
1985 (x2)	300 tons		2,400 tons

Proportional Consumption of Diosgenin Equivalent

	Corticosteroids and Spironolactone	Contraceptive Steroids
Present	94.6%	5.4%
1985	92.0%	8.0%

equivalent of 700 tons came from sources outside of Mexico but these alternate sources, raw materials and facilities were also strained to full capacity. The effects of the new processes previously mentioned have not yet been felt (10).

If, as expected, the world needs expand to twice the present demand by 1985, it is essential that sufficient capacity should be available. It is certainly not too soon to plan now and start building for the future.

After the world steroid industry has expanded in the ways which have been described, there should be sufficient capacity and low enough costs to amply provide steroid drugs for the ever-growing numbers of patients who will need them.

For the immediate present there are the prospects of higher prices and some shortages, especially in the case of corticoids, which consume such a great per cent of steroid raw materials. Reference to Tables 2, 6 and 7 can demonstrate that Mexican raw materials are no longer competitive with the newer ones that have now been put into commercial use. In fact, even if the Mexican government reverses its policy and returns to prices prevalent some years ago, it no longer seems possible that diosgenin can be considered a material of choice for 19 nor contraceptives, estrogens or diuretics.

The newer processes and raw materials for steroid drug production are now so firmly established that Mexican dioscorea may well become known as a crop that failed.

TABLE 6

*Dry Dioscorea and Diosgenin Costs**
(U.S. Currency)

	1970	1971	1972	1973	1974	1975	1976	Cultivated
Dioscorea/ton	360.00	395.00	415.00	475.00	810.00	1,622.00	5,600.00	1,258.00
Diosgenin/kilo	11.25	12.45	14.52	17.51	27.68	54.78	152.20	43.65

For 1977, the Mexican government has made a firm offer to supply dioscorea at a price which will yield diosgenin at about $95.00 per kilo.

*Source: Miller, R., Steromex S. A., Mexico, 1976 Personal communication.

TABLE 7*

Supplies of Steroidal Products and a Comparison to the Corresponding Mexican Products

Name & Location of Producer	Raw Material	Product Offered	$/Kilo Price Offered	$/Kilo[a] Price of Same Mexican Product	Use
Upjohn (U.S.)	Stigmasterol	Progesterone	250	350	Mfg. of corticoids
Upjohn (U.S.)	Stigmasterol	17-hydroxy progesterone	310	875	Progestagen
Roussel (France)	Total synthesis	Norandrostene-dione	1,000	1,400	19-norprogestagens
Searle (U.S.)	Sitosterol	Androstene-dione	135	500	Androgens and diuretics
Schering (West Germany)	Sitosterol	Androstadiene-dione	325	700	Estrogens
Jenapharm (East Germany)	Total synthesis	Diverse	---	Approximately 50% higher	Estrogens and progestagens

[a]Based upon 1975 cost of $54.78/K diosgenin.
*Source: Miller, R., Steromex, S.A., Mexico, 1976.

162

III. REFERENCES

1. An excellent biography of Russell E. Marker can be found
    in   (a) Lehmann, F. et al, *J. Chem. Educ.* 50, 195 (1973).
    Fragmentary accounts are also in   (b) Applezweig, N.,
    "Steroid Drugs", McGraw-Hill, New York, 1962.   (c) Fieser,
    L. F. and Fieser, M. "Steroids", Reinhold, New York,
    1959.
2. Marker, R. E., and Rohrmann, E. J. *J. Am. Chem. Soc.* 61,
    846 (1939).
3. Marker, R. E., Wagner, R. B., Ulschafer, P. R.,
    Wittbecker, E. L., Goldsmith, D. P. J. and Ruof, C. H.
    J. Am. Chem. Soc. 69, 2167 (1947).
4. Peterson, D. H., Murray, H. C., Eppstein, S. H., Reineke,
    L. M., Weintraub, A., Meister, P. D. and Leigh, H. M. *J.
    Am. Chem. Soc.* 74, 5933 (1952).
5. Campbell, J. A., Shepherd, D. A., Johnson, B. A. and Ott,
    A. D. *J. Am. Chem. Soc.* 79, 1127 (1957).
6. Elks, J., Phillipps, G. H., Walker, T. and Wyman, L. J.
    J. Chem. Soc. (London) 4330 (1956).
7. Djerassi, C., Ringold, H. J. and Rosenkranz, G. *J. Am.
    Chem. Soc.* 73, 5513 (1951).
8. Applezweig, N. *Chemical Week*, May 17, 1969.
9. Hardman, R. "34th International Congress of the Pharma-
    ceutical Sciences of the Federation Internationale Phar-
    maceutique," p. 60 Excerpta Medica, 1974.
10. Applezweig, N. *Chemical Week*, July 10 (1974).
11. Source material from Norman Applezweig Associates, New
    York, New York.

PLANT DERIVATIVES FOR INSECT CONTROL

Robert L. Metcalf

Department of Entomology
University of Illinois
Urbana, Illinois 61801

The employment of toxic principles of plants for insect
control probably originated in prehistoric times. Natives of
Central America have used for centuries the ground seeds of
Schoenocaulon officinale, which contain veratrine alkaloids as
louse powders. Rotenoids from *Lonchocarpus* and *Derris* have
been used as fish poisons by natives of Amazonia and Malaysia
for hundreds of years and ground roots and stems were recommended against leaf-eating caterpillars in 1849. Teas of
Nicotiana and tobacco dust containing nicotine were among the
first materials to be used as insecticides and were recommended to control aphids as early as 1763. Pyrethroids from
Chrysanthemum were reported as traveler's companions to control fleas and bedbugs before 1800. The Chinese have used
large quantities of similar indigenous plants with insecticidal properties for thousands of years.

This brief review will survey the current status of commercially important plant derived insecticides and will discuss structural optimization of plant toxicants into a variety
of useful synthetic insecticides.

I.   IMPORTANT PLANT DERIVED INSECTICIDES

A.   Pyrethroids

These active ingredients are found in the flowers of
Chrysanthemum cinerariaefolium, family Compositae and are
grown commercially in Kenya, Tanzania, Rwanda, Ecuador, and
Zaire. World production is estimated as 42,000,000 lb. dry
flowers (1972) or about 200 metric tons of active ingredients,
63% from Kenya and 22.8% from Tanzania. U.S. imports in 1972
were 108,448 lb. of flowers, (value $49,187) and 850,196 lb.
of extract (value $8,655,234) (1).

   The natural pyrethroids are used as insecticides in three
major ways: (a) as the rapid "knockdown" agents in household
sprays and aerosols, usually at 0.05-0.2% pyrethrins plus 5 to
10-fold synergist such a piperonyl butoxide or N-(2-ethylhexyl)
-bicyclo-(2.2.1)-5-heptene-2,3-dicarboximide,   (b) as livestock and cattle sprays and   (c) as sprays or dusts for mills
and warehouses, as dusts for protection of stored grains at
0.05% pyrethrins and 0.8% piperonyl butoxide, and as impreg-
nants for bags at 1:10 with piperonyl butoxide at 10 mg per
sq. ft. of cloth (2).
   The flowers of C. cinerariaefolium contain an average of
1.3% pyrenthrins which are concentrated in the achenes which
contain about 90% of the pyrethrins. The active principles
are readily extracted by solvents such as petroleum ether,
methyl alcohol, or acetone and the concentrates can be freed
from waxes and purified to 90-100% by extraction with nitro-
methane and adsorption on activated carbon.

1. *Chemistry*

   The natural pyrethrins are a mixture of 6 esters of the
two cyclopropanecarboxylic acids, chrysanthemum monocarboxylic
acid (chrysanthemic acid) and chrysanthemum dicarboxylic acid
(pyrethric acid) and three keto-alcohols pyrethrolone, cinerolone, and jasmolone (Figure 1). There are two asymmetric carbons, one in acid and one in alcohol, and the natural products
contain *d-trans*-acids and *d-cis*-esters.

2. *Toxicology*

   The pyrethroid esters differ considerably in insecticidal
potency among themselves and to various insect species. The
most careful evaluation is probably that of Sawicki et al. (3)
using topical application to the house fly *Musca domestica*:

	topical LD_{50} micrograms per female house fly
pyrethrum extract	0.56
pyrethrin I	0.59
pyrethrin II	0.38
cinerin I	1.42
cinerin II	1.14

3. *Selectivity*

   The high degree of toxicity of the pyrethrins to many
insects and their generally low toxicity to higher animals is
one of their most valuable assets. Thus the oral LD_{50} of the
pyrethrins: to the rat is 580 mg per kg and the topical LD_{50}
to the house fly is 15 mg per kg. The mammalian selectivity

rotenone

pyrethrin I

nicotine

ryanodine

cevadine

FIGURE 1

167

ratio of these two LD_{50} values is 38 (4).

Exhaustive metabolic studies have shown that the pyrethrins are degraded *in vivo* largely by mixed function oxidase attack on the methyl groups of the isopropenyl moiety of the cyclopropane carboxylic acids to form alcohols and acids, and upon the terminal C=C of the unsaturated side chains of the alcohols to form diols. These processes appear to occur relatively more efficiently in mammals than in insects (5).

4. *Mode of Action*

Little is known of the molecular mode of action of these important insecticides. However, the target site is the membrane of the nerve axon where the pyrethroids interfere with permeability to sodium and potassium ions (6).

B. Quassia

The chips of two trees *Picrasma excelsa* of Jamaica and *Quassia amara* of Surinam (Family Simarubaceae) have been used as insecticides since about 1850 (2). Teas made by steeping the chips in water are effective contact insecticides against aphids and sawflies and as much as one million lb. of chips were imported annually into the U.S. in the 1940's (7). The toxic principle found to 0.2% in the chips, is the amaroid quassin (Figure 1).

C. Ryanodine

The ground roots and stems of *Ryania speciosa* family Flacourtiaceae, of South America contain 0.1-0.2% of ryanodine alkaloid (Figure 1) that has been used commercially to control the codling moth, European corn borer and other chewing insects.

D. Sabadilla and Hellebore

The veratrine alkaloids of the roots of *Veratrum alba* and *V. viride* (hellebores) have been used as insecticides for more than 100 years but are now of little importance. The seeds of a related plant *Schoenocaulon officinale* (sabadilla) family Liliaceae, from South and Central America contain a related group of veratrine alkaloids of which cevadine (Figure 1) is the most active principle. The veratine alkaloids are present in ground seeds to 2-4%. By 1946 as much as 120,000 lb. of ground sabadilla seeds were used to control thrips and other insects in the U.S.

E. Rotenoid

Rotenoids occur in the roots and seeds of more than 68 species of plants including 21 species of *Tephrosia*, 12 of *Derris*, 12 of *Lonchocarpus*, 10 of *Milletia*, and several of *Mundulea*. The important varieties for the production of insecticides are *D. elliptica* and *D. malaccensis* from Malaysia and *L. utilis* and *L. urucu* from South America (2). U.S. import of rotenoids in 1972 was 548,875 lb. of whole roots (value $60,353) and 919,641 lb. of powdered roots (value $196,792) (1). World wide consumption of purified rotenone is reportedly 10,000-20,000 kg yearly (8).

Rotenoids are used for insecticides (a) as dusts and water dispersible powders of ground roots used to control a variety of chewing insects, especially on food plants just prior to harvest, and (b) as extracts of ground roots emulsified or in ointments for the control of cattle grubs or ox warbles encysted in the backs of cattle.

Rotenoid bearing roots contain 5 to 9% rotenoid and up to 31% ether extractives in *D. elliptica* and 8 to 11% rotenone and 25% extractives in *L. utilis*.

A. *Chemistry*

Rotenone a relatively complex chromane derivative (Figure 1), is the outstandingly active ingredient in ground roots and is from 5-10 fold more toxic than the other rotenoids. These include *elliptone* (which has a furan ring in place of ring E of rotenone, *summatrol* (15-hydroxyrotenone), malacol (15-hydroxyelliptone), *l-α-toxarol* (15-hydroxy, with a 2,2-dimethyldihydropyrone in place of ring E), and *deguelin* (H in place of 15-OH of toxicarol). Rotenone is rapidly degraded by the action of light and air to non-toxic products, primarily by hydroxylation and subsequent double bond formation between rings B and C and this provides the basis for its usefulness on edible produce within 7-10 days of harvest.

B. *Selectivity*

Although rotenone is generally considered a highly selective insecticide, the pure compound is moderately toxic to the rat, oral LD_{50} 132 mg per kg. The rapid inactivation of light and air is responsible for its unusual safety. *In vivo* rotenone is detoxified through microsomal oxidations through O-demethylation, hydroxylation of the isopropenyl side chain, and hydroxylation of the isopropenyl side chain, and hydroxylation between rings B and C (9).

C. *Mode of Action*

Rotenone has a characteristic and highly specific effect

in decreasing insect respiration. This results from specific
blocking of NADH oxidase (10). Poisoned insects gradually
become inactive and die slowly over a period of several day.

F. Nicotinoids

A variety of relatively simple alkaloids occur in the
leaves and stems of tobacco Nicotiana tabaccum and N. rustica.
Nicotine or 1-1-methyl-2-(3'-pyridyl) pyrrolidine (Figure 1)
is the principle alkaloid of insecticidal importance comprising about 97% of the total alkaloids. From the insecticidal
standpoint only two other alkaloids nornicotine or 1-2-(3'-
pyridyl)-pyrrolidine and anabasine or 1-2-(3'-pyridyl)
piperidine are of value. Nicotine occurs in N. tabaccum to
about 2-5% in the leaves and in N. rustica to 5-15%. However,
nornicotine comprises 95% of the alkaloidal content of N.
sylvestris where it occurs to about 1% in the leaf; and
anabasine from 1-2% in Anabasis aphylla, a small woody perennial of Central Asia, Asia Minor, and North Africa.

In 1972 U.S. imports of nicotine were 179,839 lb. valued
at $665,442 (1). World production is estimated at 1,250,000
lb. of nicotine sulfate and 150,000 lb. of nicotine alkaloid
(11).

Nicotine can readily be extracted from ground tobacco by
treatment with alkali followed by steam distillation or by
organic solvents. It is used insecticidally as the free base
standardized at 40% or as the less volatile and safer nicotine
sulfate standardized at 40%. Nicotine is used as a spray for
aphids and other soft-bodied insects and sometimes as a dust.
Alkaline materials such as soaps, lime or ammonium hydroxide
are often added to preparations of nicotine sulfate to liber-
ate the more toxic free nicotine base.

A. Toxicology

The relative toxicity of the nicotine alkaloids to insects is roughly nicotine 1, nornicotine 2, and anabasine 10
(12). Nicotine is highly toxic to man and higher animals and
the oral LD_{50} to the rat is 30 mg per kg and the dermal LD_{50}
to the rabbit is 50 mg per kg.

B. Mode of Action

The nicotine alkaloids affect the ganglia of the insect
central nervous system, facilitating trans-synaptic conduction
at low concentrations and blocking conduction at higher levels.
There is a pronounced structural resemblance between nicotine
and the neurohormone acetyl choline and it appears that the
protonated form of nicotine is the physiological one which in-
teracts with the postsynaptic acetyl choline receptor (12).

II. BOTANICAL INSECTICIDES AS MODELS FOR STRUCTURAL OPTIMIZA-
TION

Perhaps the most important role of insecticides derived
from plants has been as models for structural optimization
leading to new synthetic organic insecticides. The structural
optimization process has been particularly successful using
pyrethrins I as a model (4,13) and in the development of the
N-methylcarbamate insecticides which resulted from the optimization of the alkaloid physostigmine or eserine from the
calabar bean *Physostigma venenosum* (14,15).

A. Synthetic Pyrethroids

Since the elucidation of the chemical structures of the
natural pyrethrins by Staudinger and Ruzicka (16) and its further elaboration by La Forge and Barthel (17) and others, an
immense amount of effort has been directed at producing synthetic pyrethroid insecticides. The first successful product
was allethrin in which (±)-*cis-trans*-chrysanthemic acid was
esterified with a synthetic keto-alcohol differing from pyrethrolone and cinerolone in having an alkyl side chain in the
2-position (Figure 1). This compound has been used almost
entirely as a household insecticide and in pyrotechnic mosquito coils. Esterification of chrysanthemic acid with substituted benzyl alcohols led to the development of 2,4-dimenthyl-
benzyl chrysanthemate (dimethrin) (18). This compound is outstandingly safe to mammals (Table 1) and has been used in a
limited way as a mosquito larvicide safe for use in potable
water. The outstanding work of Elliott and coworkers (4,13)
has led to the development of 5-benzyl-3-furylmethyl chrysanthemate (resmethrin) which is very effective against many insects and has an outstanding MSR (Table 1) but is of short
persistence. This pyrethroid, using (+)-*trans*-chrysanthemic
acid produced by microbial fermentation (19) (bioresmethrin)
is about 2.2X as toxic to the fly and 0.18 as toxic to the rat
as resmethrin (Table 1). In resmethrin, the spatial configuration of the phenyl group is very similar to that of the
cis-butadienyl side chain of allethrolone. Enhanced insecticidal activity results from the greater resistance of the
phenyl group to *in vivo* detoxication in the insect. The replacement of the asymmetric alcohol (+)-pyrethrolone by the
symmetrical 5-benzyl-3-furylmethyl alcohol has substantially
simplified synthesis.

A more recent modification of the alcoholic component is
3-phenoxybenzyl (±)-*cis-trans*-chrysanthemate (phenothrin)
which also preserves the essential configuration of an unsaturated center in a position corresponding to the side chain
of pyrethrin I. This pyrethroid is substantially less susceptible to degradation by light and air and is much more per-

sistent than resmethrin. Still further enhancement of insecticidal activity resulted from incorporation of an α-CN group
into the alcohol moiety. Thus α-cyano-3-phenoxybenzyl (±)-*cis*-
2,2-(dibromovinyl)-1,1-dimethylcyclopropane carboxylate (NRDC-
161) is about 2X as toxic to the house fly as bioresmethrin
(Table 1) and seems to be the most active synthetic pyrethroid
yet devised.

Other types of synthetic pyrethroids have involved modifications of the stryctyre of chrysanthemic acid and esterification with the synthetic alcohol moieties discussed above.
Particularly interesting pyrethroids have been obtained from
the dichloro- and dibromovinyl esters of chrysanthemic acid
where the (+)-*cis*-acids seem to be more active than the *trans*-
acids. The (+)-*cis*,*trans*-2,2-dichlorovinyl-1,1-dimethyl-
cyclopropane carboxylic acid ester of 3-phenoxybenzyl alcohol
(permethrin) is outstandingly active and is of greatly enhanced environmental stability so that its residual activity is
measured in weeks.

The most recent modifications of the pyrethroid nucleus
involve replacement of the chrysanthemic acid moiety with acids
such as α-isopropyl-*p*-chlorophenylacetic acid. When this is
esterified with α-cyano-3-phenoxybenzyl alcohol the resulting
pyrethroid (Table 1) is a highly effective and relatively persistent insecticide. Structural optimization has now progres-
sed so far that only the essential stereochemical configuration of a few essential atoms has been preserved.

B.   N-Methylcarbamates

The prinicple alkaloid of the seeds of the calabar bean
Physostigma venenosum, physostigmine or eserine was used as an
ordeal poison in W. Africa for centuries and was recognized as
a miotic as early as 1862. Stedman and Barger (20) elucidated
its structure as the N-methylcarbamate of a trimethyl indole
derivative (Figure 1). Physostigmine was found to be a strong
inhibitor of the enzyme acetylcholinesterase and has been used
in medicine for the treatment of glaucoma and myosthemia
gravis. A number of synthetic derivatives, especially
prostigmine or *m*-trimethylammoniumphenyl N,N-dimethylcarbamate,
were developed as pharmaceuticals but the carbamates were
uninvestigated as insecticides until the studies of Gysin (21)
and Kolbezen et al. (22).

Although physostigmine has insecticidal activity, the
quaternary ammonium derivatives are totally inactive. Kolbezen
et al. (22) showed that neutral lipophilic phenyl N-methylcarbamates had substantial insecticidal properties. Many hundreds of substituted phenyl N-methylcarbamates have been evaluated as insecticides (14,15) and activity has been demonstrated to depend upon complementarity of structure to acetyl

TABLE 1

Toxicity and Selectivity of Pyrethroids

		topical LD$_{50}$	oral	
		Musca domestica	rat	MSR[a]
R	R'	mg per kg	mg per kg	
(+)-*trans*-(CH3)2C=CH	2-pentadienylcyclopentenolone (pyrethrin I)	15	580	38
(+)-*cis-trans*-(CH3)2C=CH	2-allylcyclopentenolone (allethrin)	10	770	77
(+)-*cis-trans*-(CH3)2C=CH	CH2C6H3(CH3)2-(2,4) (dimethrin)	50	10,000	200
(+)-*cis-trans*-(CH3)2C=CH	CH2C4H2OCH2C6H5-(3) (resmethrin)	0.6	1,400	2,300
(+)-*trans*-(CH3)2C=CH	CH2C4H2OC6H5-(3) (bioresmethrin)	0.25	8,000	32,000
(+)-*cis-trans*-((CH3)2C=CH	CH2C6H4OC6H5-(3) (phenothrin)	1.1	5,000 (mouse)	5,000
(+)-*cis-trans*-Cl2C=CH	CH2C6H4OC6H5-(3) (permethrin)	1.35	600	440
(+)-*cis*-Br2C=CH	CH(CN)C6H4OC6H5-(3) (NRDC-161)	0.17	128 / 33 (mouse)	880
Cl-C6H4CH(CHMe2)CO (acid component)	CH(CN)C6H4OC6H5-(3)	4.0	200 (mouse)	50

[a] Mammalian selectivity ratio – rat LD$_{50}$/fly LD$_{50}$.

TABLE 2

Toxicity and Selectivity of Methyl Carbamates

$$\overset{\text{O}}{\overset{\|}{R}}\text{OCNHCH}_3$$

$R =$		topical LD_{50} *Musca domestica* mg per kg	oral LD_{50} rat mg per kg	$MSR^{[a]}$
Me … N–Me Me (N)	physostigmine	ca 100	3 (mouse)	0.03
		105	150	1.43
phenyl–O–i–Pr	propoxur	25.5	116	4.55
phenyl (t–Bu, t–Bu)	butacarb	39	>4,000	>103
naphthyl	carbaryl	900	540	0.60

Me Me	carbofuran	4.6	4.0	0.87

Me │ $CH_3SCCH=N-$ Me	aldicarb	5.5	0.93	0.17
$CH_3SCH=N-$ Me	methomyl	10.5	17	1.62

[a]Mammalian selectivity ratio = rat LD_{50}/fly LD_{50}.

175

choline. Thus these insecticides act as synthetic substrates
for the acetylcholinesterase enzyme but possess very slow turn-
over numbers and carbamylate the active site.

Several of the synthetic aryl N-methylcarbamates are important insecticides (Table 2). Carbaryl is the insecticide
in largest use in the United States, largely because of its
relative safety and biodegradability, with an annual production of about 80 million lb. Carbofuran despite its high toxicity to higher animals is a very widely used soil insecticide.
Propoxur is a very effective household insecticide and a suc-
cessor to DDT for control of malaria mosquitoes. Aldicarb or
2-methyl-2-methylthiopropionaldehyde oxime N-methylcarbamate
and methomyl or S-methyl-N-methylcarbamoyloxy thioacetimidate
represent newer types of insecticidal carbamates which are
N-methylcarbamoyloximes (=NOC(O)NHCH$_3$). Despite their highly
optimized structures, they preserve the essential steric
characteristics of physostigmine which enable them to serve as
substrates for acetyl cholineserase producing inhibition
through carbamylation of the serine hydroxyl at the active
site. These oxime carbamates are very active insecticides but
have a low degree of mammalian selectivity (Table 2). As
shown in Table 2, the greatest problem with the carbamates is
to produce compounds with a high degree of mammalian safety
and selectivity. Butacarb, used in Australia for the control
of wool maggots or sheep blowfly, is a notable exception.

III. REFERENCES

1. "Pesticide Review", U.S. Department of Agriculture. Agri-
    cultural Stabilization and Conservation Service, Washing-
    ton, D.C. 1973.
2. Metcalf, R. L., Metcalf, C. L. and Flint, W. P., "Destruc-
    tive and Useful Insects" McGraw Hill, New York 1962.
3. Sawicki, R. M., Elliott, M., Gous, J. G., Snarey, M.,
    Thain, E. M. J. Sci. Food Agr., 1962 (3), 172.
4. Elliott, M. Bull. World Health Org. 44, 315 (1971).
5. Hollingworth, R. M. in "Insecticide Biochemistry and Phy-
    siology", (C. F. Wilkinson, Ed.) Chap. 12, Plenum Press,
    New York 1976.
6. Narahashi, T. Bull. World Health Org. 44, 337 (1971).
7. Jacobsen, M. and Crosby, D. G. Eds., "Naturally Occurring
    Insecticides", M. Dekker, New York 1970.
8. Fishbein, L. M. in "Insecticide Biochemistry and Physiol-
    ogy", (C. F. Wilkinson, Ed.) Chap. 14, Plenum Press, New
    York 1976.
9. Nakatsugawa, T. and Morelli, M. A. in "Insecticide Bio-
    chemistry and Physiology", (C. F. Wilkinson Ed.) Chap. 2,
    Plenum Press, New York 1976.
10. Fukami, J. in "Insecticide Biochemistry and Physiology",

(C. F. Wilkinson, Ed.) Chap. 10, Plenum Press, New York 1976.

11. Schmeltz, I. in "Naturally Occurring Insecticides", (M. Jacobs and D. G. Crosby, Ed.) M. Dekker, New York 1971.

12. Yamamoto, I. in "Advances in Pest Control Research", Vol. VI, (R. L. Metcalf, Ed.) pp.231-260, Interscience, New York 1965.

13. Elliott, M. in "The Future for Insecticides: Needs and Prospects", (R. L. Metcalf and C. M. McKelvey, Jr., Eds.) pp. 163-190, Wiley-Interscience, New York 1976.

14. Metcalf, R. L. and Fukuto, T. R. *J. Agr. Food Chem.* 13, 220 (1965).

15. Metcalf, R. L. *Bull. World Health Org.* 44, 43 (1971).

16. Staudinger, H. and Ruzicka, L. *Helv. Chim. Acta* 7, 448 (1924).

17. LaForge, F. B. and Barthel, W. F. *J. Org. Chem.* 10, 222 (1945).

18. Barthel, W. F. in "Advances in Pest Control Research", Vol. IV, (R. L. Metcalf, Ed.) pp. 34-74, Interscience, New York 1961.

19. Rauch, F., Lhoste, J. and Berg, M. L. *Mededelingen Fakulteit Landbouwwetenschappen Gent* 37 (2), 755 (1972).

20. Stedman, E. and Barger, G. *J. Chem. Soc.* 127, 247 (1925).

21. Gysin, J. *Chimia* 8, 205 (1954).

22. Kolbezen, M., Metcalf, R. L. and Fukuto, T. R. *J. Agr. Food Chem.* 2, 864 (1954).

EVOLUTIONARY DYNAMICS OF SORGHUM DOMESTICATION

J. M. J. de Wet

Crop Evolution Laboratory
Department of Agronomy
University of Illinois
Urbana, Illinois 61801

and

Y. Shechter

Department of Biological Sciences
Lehman College
Bronx, New York 10468

Archaeological evidence suggests that wild grasses were an important food source of proto-agriculturalists. Some 60 species are still regularly harvested in Africa by hunting-gathering as well as agricultural people (1-3). Only eight of these became fully domesticated. Evidently, harvesting alone does not lead to domestication. Sorghum (*Sorghum bicolor* (Linn.) Moench) is botanically the best known native African cereal. It is widely grown across Africa and southern Asia (4), and the wild races of *S. bicolor* are still harvested for human food in the African savanna. Changes in phenotype and adaptation brought about by domestication will be discussed.

I. TAXONOMY

The genus *Sorghum* Moench is widely distributed in the warmer regions of the Old World (5-7), with section *Sorghastrum* also extending into the New World. It is an immensely variable genus, and is usually subdivided into sections *Chaetosorghum*, *Heterosorghum*, *Parasorghum*, *Sorghastrum*, *Sorghum* and *Stiposorghum* (8, 9). Phylogenetic affinities among these sections are poorly understood. However, it is well established that all cultivated grain sorghums and their

closest wild relatives belong in section *Sorghum* (4, 10).
Section *Sorghum* has been variously classified by taxonomists
(5, 6), plant breeders (11-12), and crop evolutionists (13).
The classification presented in Table 1 is based on suggestions by Harlan and de Wet (14). Two species are recognized;
S. halepense (Linn.) Pers. to include tetraploid (2n=40)
rhizomatous taxa, and *S. bicolor* to include all diploid (2n=
20) taxa classified by Snowden in his section *Eu-Sorghum*.

Sorghum halepense is a perennial with well developed
rhizomes. It is characterized by a mediterranean ecotype
with narrow leaves and slender culms, and a more robust tropical ecotype. The tropical ecotype extends from southern
India to western Pakistan where it is replaced by the mediterranean ecotype which is widely distributed in the Near East,
southern Europe and coastal North Africa. The mediterranean
ecotype was introduced to the New World around 1820 where it
is now often encountered as a roadside weed. It crosses readily with cultivated sorghums and derivatives of such hybridization are known as *S. almum* Parodi in South America and
Johnsongrass in North America. *Sorghum halepense* did not contribute directly towards the origin of domesticated sorghum.
Its distribution falls outside the major range of sorghum cultivation in Africa.

Sorghum bicolor is divided into subspecies *bicolor* and
subspecies *arundinaceum* (Desv.) de Wet and Harlan. Cultivated
grain sorghums are included in ssp. *bicolor* and all spontaneous taxa in ssp. *arundinaceum* (Table 1).

Subspecies *arundinaceum* is divided into races arundinaceum, aethiopicum, drummondii, propiquum, virgatum, and verticilliflorum.

Race propiquum is the only member of the subspecies not
confined to Africa. It is widely distributed in tropical
southern Asia. Propiquum resembles *S. halepense* in being a
rhizomatous perennial but differs from it in having smaller
grains and larger inflorescences. It crosses readily with
cultivated sorghums and hybrids between them have been collected as weeds in cultivated sorghum fields across southern
Asia.

Race aethiopicum is widely distributed along the northern
edge of the savanna from western Ethiopia to central Mauritania. It is characterized by relatively small inflorescences
with somewhat erect branches and large spikelets. Race verticilliflorum is recognized to include several Snowdenian
species and derivatives of hybridization among them (10). It
is characterized by large open inflorescences with branches
that are divided along their length. This race is the common
wild sorghum of the African savanna from southern Africa to
eastern Nigeria. Where verticilliflorum and aethiopicum are

TABLE 1

Classification of Sorghum section Sorghum

Taxonomy	Taxa included
S. halepense (Linn.) Pers.	S. almum Parodi
	S. controversum (Steud.) Snowden
	S. miliaceum (Roxb.) Snowden
S. bicolor (Linn.) Moench	
ssp. *arundinaceum* (Desv.)	
de Wet et Harlan	
race arundinaceum	S. arundinaceum (Desv.) Stapf
	S. vogelianum (Piper) Stapf
race aethiopicum	S. aethiopicum (Hack.) Rupr. ex
	Stapf
race drummondii	S. aterrimum Stapf
	S. drummondii (Steud.) Millsp.
	Chase
	S. elliottii Stapf
	S. hewisonii (Piper) Longley
	S. niloticum (Stapf ex Piper)
	Snowden
	S. nitens (Bussey ex Pilg.)
	Snowden
	S. sudanense (Piper) Stapf
race propiquum	S. propiquum (Kunth) Hitchc.
race verticilliflorum	S. brevicarinatum Snowden
	S. castaneum Hubb. et Snowden
	S. lanceolatum Stapf
	S. macrochaeta Snowden
	S. panicoides Stapf
	S. pugionifolium Snowden
	S. somaliense Snowden
	S. usambarense Snowden
	S. verticilliflorum (Steud.)
	Stapf
race virgatum	S. virgatum (Hack.) Stapf
ssp. *bicolor* (Linn.) Moench	
race bicolor	S. bicolor (Linn.) Moench
	S. dochna (Forsk.) Snowden
	S. exsertum Snowden
	S. nervosum Bess. ex Schult.
	(in part)
	S. splendidum (Hack.) Snowden
race caudatum	S. caudatum Stapf (in part)

Table 1 - continued

Taxonomy	Taxa included
	S. nigricans (R. et P.) Snowden (in part)
race durra	S. cernuum Host.
	S. durra Stapf
race guinea	S. conspicuum Snowden
	S. gambicum Snowden
	S. guineense Stapf
	S. margaritiferum Stapf
race kafir	S. caffrorum Beauv.
	S. coriaceum Snowden
caudatum-bicolor	S. nervosum Bess. ex Schult. (in part)
	S. notabile Snowden (in part)
durra-bicolor	S. ankolib Stapf
	S. rigidum Snowden
	S. subglabrescens Schweinf. et Aschers.
guinea-caudatum	S. caudatum Stapf (in part)
	S. dulcicaule Snowden
	S. elegans (Koern.) Snowden (in part)
	S. notabile Snowden (in part)
guinea-durra	Not recognized by Snowden
guinea-kafir	S. roxburghii Stapf
guinea-bicolor	S. melaleucum Stapf
	S. mellitum Snowden
durra-caudatum	S. caudatum Stapf (in part)
kafir-bicolor	S. basutorum Snowden
	S. elegans (Koern.) Snowden (in part)
	S. miliiforme (Hack.) Snowden
	S. nervosum Bess. ex Schult. (in part)
	S. simulans Snowden
kafir-caudatum	S. nigricans (R. et P.) Snowden (in part)
kafir-durra	Not recognized by Snowden
various races with large glumes due to py allele	S. membranaceum Chiov.

sympatric they hybridize and grade morphologically into each
other. Both races are locally abundant and frequently harvested for human food. The southern Nguni considers verticilliflorum to be the progenitor of cultivated sorghums.

Race arundinaceum is a forest grass of the Guinea coast
and Congo. It is characterized by large open inflorescences
with pendulous branches that are undivided near the base.
Along the edge of the tropical forest arundinaceum grades morphologically into verticilliflorum. Its forest habitat excludes arundinaceum as a possible progenitor of cultivated
sorghums.

Race virgatum is weedy along irrigation ditches and stream
banks in central Sudan, and extends along the Nile Valley to
Cairo. It is a slender grass with small inflorescences having
relatively few branches. Distribution and low yield seem to
exclude virgatum as a possible ancestor of cultivated sorghums.

Race drummondii includes all weedy taxa that combine
characteristics of cultivated sorghums and wild members of
ssp. *arundinaceum*. They often mimic their cultivated parents
in inflorescence and spikelet morphology, but can readily be
recognized by spikelets which articulate at maturity.

Subspecies *bicolor* is extremely variable. It includes 28
cultivated Snowdenian species (Table 1). Races bicolor, caudatum, durra, guinea and kafir are recognized on the basis of
spikelet morphology. Stabilized derivatives of hybridization
among races are identified as intermediate races (13).

Race bicolor extends across the range of sorghum cultivation in Africa, but is never extensively cultivated. It is
characterized by open inflorescences and long clasping glumes
that are at least 3/4 as long as the grain. Bicolor sorghums
resemble members of drummondii except that their spikelets
are persistent. Some probably represent relics of a primitive
cultivated race, but most can be reconstructed by backcrossing
members of drummondii with their cultivated parents.

Race caudatum is commonly grown in parts of the Sudan,
Chad, Nigeria and most of Uganda. This race has very characteristic turtle-backed grains. Kinds with open inflorescences
are usually grown in wetter regions, and those with more compact inflorescences in drier regions. Caudatum is an old
race. Carbonized grains from Daima near Lake Chad that date
back to around the ninth century A.D. (32) have the characteristic turtle-back shape of present-day caudatum.

Race durra is widely distributed in the Near East and
India. In Africa it is closely associated with Islam. It is
widely grown in lowland Ethiopia and is essentially the only
sorghum cultivated along the southern fringe of the Sahara.
Spikelets of durra sorghums are broadly ovate with lower
glumes that are frequently wrinkled near the middle. Inflorescences are usually very compact, but sometimes large and

open. Distribution suggests an origin outside Africa as selections out of an early African domesticated race. A large
sheath of sorghum recently discovered in a building at Qasr
Ibrim and dated to Meroitic times (15) probably belong with
race durra or durra-bicolor.

Race guinea is characterized by glumes that gape widely at
maturity, and the grain is obliquely twisted between them.
Inflorescences are usually large and somewhat open, often with
pendulous branches at maturity. This is primarily a West
African race with a secondary center of dominance in Malawi.
It is grown in areas with as much as 120 inches of rainfall
during the growing season but is equally well adapted to regions with only one quarter that amount of rain. We have not
seen any archaeological sorghum that belongs to this race.
But, grains from Muramasapwa in Malawi dating back to the
ninth century A.D. (16) may represent guinea sorghum.

Race kafir is the most common cultivated sorghum south of
the equator. Spikelets are broadly elliptic with the glumes
tightly clasping the usually longer grain. Inflorescences
are loose to compact. This race has been found associated
with early iron-age Bantu dwelling sites from Zambia to South
Africa (16-20).

II.   PROGENITORS OF CULTIVATED SORGHUMS

Sorghum is a native African cereal. It originated from
wild members of *S. bicolor* ssp. *arundinaceum*. But, it is not
known with certainty which wild race or races were domesticated. Snowden (5) and Portères (21) proposed that aethiopicum,
arundinaceum and verticilliflorum could have been domesticated
independently giving rise to durra, guinea and kafir sorghums
respectively. However, close affinities between specific cultivated races and specific assumed wild progenitors cannot be
demonstrated experimentally (4).

Races propiquum, arundinaceum and virgatum probably did
not contribute directly towards the origin of cultivated sorghum. Propiquum is a South Asian grass, and although sorghum
is now widely grown in this region it probably was introduced
from Africa as a domesticated cereal (7, 22). Arundinaceum
is a forest grass, and although guinea sorghums are widely
grown in the tropical forest of West Africa the race is basically a savanna cereal (23). Virgatum has a narrow distribution along water courses of northeastern Africa, and although
durra sorghums are now grown in this area they probably were
introduced from the Near East and India in historical times.

Races aethiopicum and verticilliflorum are the most
likely progenitors of cultivated sorghum. Botanically they

represent the extremes of a wild sorghum complex that extends
across the African Savanna. Both are attractive wild cereals.
They often are dominant grasses in the northern savanna, yield
well, and their grains are large and usually palatable.
Either race or both could have been domesticated.

Sorghum was domesticated south of the Sahara, since it is
in the savanna where wild sorghums are common. Murdock (24)
proposed that West Africa was an area of independent agricultural invention around 5000 B.C. and that sorghum could have
been domesticated at this time (cf. also reference 25).
Doggett (7), on the other hand, suggested that domestication
took place some 5000 years ago somewhere in the northeastern
quadrant of Africa. Present-day distribution of wild sorghums
seem to rule out both regions as likely centers of sorghum
domestication, and Harlan (26) suggested a diffused center
somewhere in the savanna between Sudan and Mauritania. Wild
sorghums belonging to the aethiopicum-verticilliflorum complex
are still abundantly available for harvesting in this region,
and also provide excellent grazing to nomadic pastoralists.

III. THE DOMESTICATION PROCESS

Among several dozen grasses commonly harvested by pastor-
alist in the African savanna only sorghum (*Sorghum bicolor*)
pearl millet (*Pennisetum americanum* (Linn.) K. Schum.), finger
millet (*Eleusine corocana* Gaertn.), African rice (*Oryza glaberrima* Steud.), fonio (*Digitaria exilis* Stapf), ibura
(*Digitaria iburu* Stapf), West African finger millet (*Brachia-
ria deflexa* (Schum.) C. E. Hubbard), and tef (*Eragrostis tef*
(Zucc.) Trotter) ever became fully domesticated.

It is quite obvious that harvesting alone does not necessarily lead to domestication, and that all wild cereals do
not necessarily lend themselves to domestication.

The evolutionary dynamics of plant domestication are not
easy to study experimentally. All cereals were domesticated
by early cultivators, probably before the beginning of the
Christian era. The archaeological record suggests that hundreds or perhaps thousands of generations of sowing were required to produce the domesticated cereals as we know them
today. Why some wild cereals became domesticated, while other
equally promising ones did not, is difficult to decide. Grasses that were selected for domestication must have yielded
well as wild cereals, must have been well regarded as a
source of food, and perhaps most important of all, must have
lent themselves to domestication.

The initial step in domestication is a change in adaptation from a natural, relatively undisturbed habitat, to one

that is permanently and continuously being disturbed by man. Although harvesting is certainly a form of disturbance, it will not lead to domestication. It is the seeds that were not harvested that will produce the next generation. Thus, although harvesting will change gene frequencies in populations, selection pressures associated with this process will reinforce wild-type adaptations. In the case of wild cereals, adaptations for natural seed dispersal will automatically be selected for, and habitat adaptation will be maintained. Only when harvested seeds are planted in specially prepared habitats, generation after generation, is the true domestication process initiated. Evolution of the population now becomes directed by a completely different set of selection pressures (cf. reference 27). Adaptations associated with an increase in percentage grain recovered through harvesting, and an increase in survival potential under conditions of continuous human disturbance of the habitat, now will have selective advantages over alternative wild-type adaptations.

Two types of plants successfully occupy permanently disturbed man-made habitats, crops and weeds. The ecological differences between them are often difficult to define (27, 28). In general, crops occupy habitats specifically created for them by man, while weeds are more aggressive and can successfully invade all kinds of permanently man-disturbed habitats. Many weedy species, such as several taxa of *Amaranthus* and *Chenopodium*, may actually be semi-domesticates. Usually, however, the weedy habits of close relatives of domesticated species were inherited from cultivated members of the same phylogenetic complex. Weeds do not give rise to domesticates, but evolve as a result of domestication or some other human activity. Populations are weedy when they compete ecologically with man for the habitat he is creating for himself, and populations are truly domesticated when they depend on man for suitable habitats to occupy.

Races of cultivated species vary considerably in degree of weediness. The wild races of domesticated cereal species, at least as we know them today, are aggressive colonizers of all kinds of disturbed habitats. But, they are truly wild, and cannot successively invade habitats that are permanently being disturbed by man. This weedy habit, however, will facilitate domestication, and the absence of such a weedy tendency may have excluded many promising wild cereals from becoming domesticated.

In the archaeological record, the first obvious morphological characteristic associated with cereal cultivation is a change from fragile to non-articulating inflorescences. Shattering of individual fruits at maturity is an absolute requirement for natural seed dispersal. Persistence of individual fruits on the inflorescence branches facilitates

harvesting and makes natural seed dispersal impossible. Tough
rachises certainly suggests cultivation but do not necessarily
indicate full domestication. Fragile rachises, as a mechanism
for natural seed dispersal, is inherited as a genetic dominant
over persistent fruits. However, a few plants with non-articulating inflorescences are frequently encountered in large
stands of wild grasses. Realizing the advantage in having
large populations of a cereal with inflorescences that remain
intact at maturity may have led man to deliberately plant such
selected seed. However, even if man started sowing wild cereals, the mere process of planting harvested seed one generation after the other will eventually result in cereal populations with uniformly tough rachises.

IV. MORPHOLOGICAL CHANGES ASSOCIATED WITH DOMESTICATION

The aethiopicum-verticilliflorum complex of *S. bicolor*
meets all prerequisites to qualify as the ancestors of domesticated grain sorghums. These wild races are widely distributed, are aggressive natural colonizers, are often abundant
over large areas, yield resonably well, their grains are palatable, and they probably formed an important food source of
protoagriculturists across the African savanna.
The archaeological record of sorghum domestication is
incomplete at this time, and phenotypic changes associated
with evolution under domestication have to be deduced from
comparative morphological studies of cultivated and wild sorghum races. Several morphological changes seem to be the
result of automatic selection for adaptation to cultivation,
while others are associated with conscious selection by man
for specific characteristics (Table 2).
Cultivation automatically leads to an increase in grain
size, and in sorghum also to a change in grain shape. Both
characteristics result from an increase in endosperm in relation to embryo size. Sowing greatly increases the number of
potential plants per unit area of space. The most vigorous
seedlings, those with the largest food source in the grain
will, therefore, have a selective advantage. High density of
plants will also automatically select against dormancy of
embryos, an advantageous characteristic in nature. Dormancy
in sorghum is associated primarily with substances in the
flower bracts, with the result that in cultivated sorghums,
the grains usually thrash completely free from all flower
appendages.
Reaping, and sowing harvested seed automatically leads to
selection for characteristics that will increase percentage
grains recovered. This may not actually increase total yield

TABLE 2

Morphological changes brought about
by domestication of sorghum

A. Changes associated with automatic selection for adaptation
    to man-made habitats
    1. Larger grains as a result of an increase in amount of
        endosperm
    2. Change in grain shape from elliptic to more obovate in
        front view as it increases in size
    3. Change from persistence of flower parts to free
        thrashing grains
    4. Change from fragile spikelet clusters to spikelets
        that remain attached to the rachises at maturity
    5. Contraction of inflorescence axis and branches to
        produce in extreme cases tightly compact inflores-
        cences
    6. Reduction of tillering to produce plants with often
        only a single stem
    7. Reduction in aerial branching to produce plants that
        often have a single unbranched stem with a solitary
        terminal inflorescence

B. Changes associated with conscious manipulation by man
    1. Increase in variability of plant types
    2. Increase in variability of inflorescence types
    3. Increase in variability of spikelet types
    4. Increase in variability of grain types
    5. Increase in distribution of the species

---

of individual plants in the initial stages of domestication,
but probably will rapidly lead to an increase in overall
yield per acre. Early in the domestication process, an increase in seedling vigor allows for an increase in the number
of plants that can survive per unit area of space. Crowding
will reduce tillering as well as aerial branching, with the
result that fewer inflorescences are produced per plant, but
these will mature more uniformly. These changes in plant
structure, associated with a change to nonarticulating inflorescences, will not only facilitate, but also greatly increase
the percentage grain recovered through harvesting.

     Morphological changes are also brought about by con-
scious as well as unconscious selection, and isolation, by
man. The range of the species is often greatly increased by
cultivation, giving rise to new phenotypic expressions of

ancestral heredities, as well as changes associated with selection for adaptation to the new environments. Compact inflorescences in sorghum usually indicate cultivation in com-
paratively dry regions. In the wetter parts such inflorescences are subject to fungus infection and consequently sorghums
grown in the broad-leaf savanna usually have large inflorescences with spreading branches. Migration of crops also leads
to isolation, often of population that originated from small
selected seed samples. Morphological pecularities of such
small isolated populations can thus become fixed (the concept
of "genetic drift" as discussed by Sewall Wright (29). This
probably is the basis for racial differentiation in cultivated sorghum (30).

Specific selection by man for different uses greatly increased variability of the stature and structure of cultivated
grain sorghums (cf. reference 5). Sorghum is an extremely
versatile crop (31, 32). The grain is eaten raw when slightly
immature. When ripe the whole grain is parched or boiled after the seedcoat is removed. After grinding the flower is
mixed with cold water as a drink, boiled as a porridge or
baked into unleavened bread. Sorghum is also grown for the
red dye produced by stem and leaf; the sweet culms of some
kinds are chewed; the bitter grains of others are used to
make malt and flavor beer; while the tall stems of many kinds
are used as fuel and building material, and for animal feed.
These numerous uses of sorghum are often mutually exclusive.
Different kinds of sorghum are usually grown for different
uses, and even adjacent villages may have different kinds for
similar uses. No wonder, therefore, that sorghum is one of
the most variable cereals.

Cultivated grain sorghums range in size from less than
3 feet to over 15 feet. The stems are pithy or juicy and
sweet. Inflorescences range from large and open with spreading or pendulous branches arranged along an elongated or
contracted central axis, to ones that are small or large and
highly compacted (cf. reference 13). Spikelet morphology is
equally variable. Glumes may completely cover the grain or
may be much shorter than the bulging, flat, concave, convex,
ovate, obovate or elliptic grain (30). Grain color varies
from white through most shades of yellow, brown, red and
purple, and may be starchy or vitreous.

V. CONCLUSIONS

Sorghum is the most important native African cereal. It
is primarily a savanna crop but is also grown in areas with
as little as 10 inches and as much as 120 inches of rainfall

during the growing season. Cultivated sorghums and their
closest wild relatives are botanically recognized as a single
species *S. bicolor* (Linn.) Moench. Cultivated sorghums are
included in subspecies *bicolor* and divided on the basis of
spikelet structure into five basic races, bicolor, caudatum,
durra, guinea and kafir, and ten intermediate races that combine characteristics of any two of these in all possible combinations. Their spontaneous relatives are included in subspecies *arundinaceum* and divided, on the basis of distribution
and inflorescence structure, into races arundinaceum, aethiopicum, drummondii, propinquum, verticilliflorum, and virgatum.
Distribution and habitat adaptation seem to exclude races
arundinaceum, propiquum and virgatum as probably progenitors
of cultivated sorghums. Races aethiopicum and verticilliflorum are savanna grasses, often abundant and are still commonly
harvested in the wild for human consumption. Either race or
both could have been domesticated at different times and
places across their ranges in the savanna. Affinities between specific kinds of wild sorghums and specific cultivated
races cannot be established on the basis of biosystematic
studies, neither can phylogenetic affinities between the cultivated races be determined with certainty. Additional
archaeological data are needed before the changes brought
about by domestication can be put in true evolutionary perspective. Comparative morphological studies, however, give
some insight into the evolutionary dynamics of the domestication process. Automatic selection towards adaptation for survival in permanently disturbed man-made habitats, conscious
selection by man for specific uses, as well as migration and
consequent isolation brought about a multitude of phenotypic
changes that can serve as indicators of cultivation in an
archaeological context.

VI. REFERENCES

1. Busson, F. "Plantes alimentaires de l'ouest African--
    Etude botanique, biologique et chimique." Leconte,
    Marseille, France, 1965.
2. Jardin, C. "List of foods used in Africa," Food Consump-
    tion and Planning Branch Nutrition Div., FAO, Rome and
    Nutrition Sect. Office of Intl. Res., Natl. Inst. of
    Health. Bethesda, Maryland, 1967.
3. Gast, M., "Alimenta des populations de l'ahaggar. Mem-
    oires 8 CRAPE," Paris, 1968.
4. de Wet, J. M. J. and Huckabay, J. P. *Evol.* 21, 787
    (1967).

5. Snowden, J. D. "The cultivated races of *Sorghum*" Allard and Son, London, 1936.
6. Snowden, J. D. *J. Linn. Soc. London* 55, 191 (1955).
7. Doggett, H. in "Crop Plant Evolution", (Sir Joseph Hutchinson, Ed.) pp. 50-69. Cambridge University Press, London, 1965.
8. Garber, E. D. *Univ. Calif. Publ. in Bot.* 23, 283 (1950).
9. Celarier, R. P. *Cytologia* 23, 395 (1959).
10. de Wet, J. M. J., Harlan, J. R., and Price, E. G. *Amer. J. Bot.* 57, 704 (1970).
11. Murty, B. R., Arundachalam, V., and Saxena, M. B. L. *Ind. J. Genet. and Plant Breed.* 27, (Supplement), 1 (1967).
12. Jakusyevsky, E. S. *Bull. Appl. Bot. Genet. and Plant Breed.* 41, 148 (1969).
13. Harlan, J. R., and de Wet, J. M. J. *Crop. Sci.* 12, 172 (1972).
14. Harlan, J. R., and de Wet, J. M. J. *Taxon* 20, 509 (1971).
15. Plumley, J. M. *J. Egypt. Archaeology* 56, 12 (1969).
16. Robinson, K. R. *South Afr. Archaeological Bull.* 21, 5 (1966).
17. Fouche, L. Mapunqubwe, ancient Bantu civilization on the Limpopo. Cambridge University Press, Cambridge, 1937.
18. Summers, R. "Inyanga", University Press, Cambridge, 1958.
19. Fagan, B. M. "Iron age cultures in Zambia", Chatto and Windus, London, 1967.
20. Phillipson, D. W., and Fagan, B. *J. Afr. Hist.* 10, 199 (1969).
21. Portères, R. *J. Afr. Hist.* 2, 195 (1962).
22. Benson, C., and Subba Rao, C. K. *Bull. Dept. of Agr. Madras* 3, 64 (1906).
23. de Wet, J. M. J., Harlan, J. R., and Kurmarohita, B. *East Afr. Agr. and For. J.* (1972).
24. Murdock, G. P. *Geog. Rev.* 50, 521 (1959).
25. Baker, H. G. *J. of Afr. Hist.* 3, 229 (1962).
26. Harlan, J. R. *Science* 174, 468 (1971).
27. Harlan, J. R., and de Wet, J. M. J. *Econ. Bot.* 19, 16 (1965).
28. Harlan, J. R., de Wet, J. M. J., and Price, E. G. *Evol.* 27, 311 (1973).
29. de Wet, J. M. J. *Proc. Okla. Acad. Sci.* 47, 14 (1968).
30. Wright, S. *Evol.* 2, 279 (1948).
31. de Wet, J. M. J., and Harlan, J. R. *Econ. Bot.* 25, 128 (1971).
32. Curtis, D. L. *Field Crops Absts.* 18, 145 (1965).
33. Connah, G. "Progress report on archaeological work in Bornu 1964-1966. Northern History Research Scheme, 2nd Int. Rept.", Zaria (1967).
34. Harlan, Jr. R. and Pasqueneau, J. *Econ. Bot.* 22, 70 (1969).

THE ORIGIN AND FUTURE OF WHEAT*

E. R. Sears

Agricultural Research Service
U. S. Department of Agriculture
and
Agronomy Department
University of Missouri
Columbia, Missouri 65201

The sources of two of the three genomes of common wheat, *Triticum aestivum* L. em. Thell., have been determined beyond reasonable doubt. Genome A came from wild diploid wheat, *T. monococcum* L., which was one parent of a cross that gave rise to wild tetraploid wheat, *T. turgidum* L. *T. turgidum* (genome formula AABB) was taken into cultivation in the Near East about 10,000 years ago and was soon converted into a cultivated (non-fragile) form (1,2). Eventually the cultivated tetraploid hybridized with the weed *T. tauschii* (Coss.) Schmal. (= *Aegilops squarrosa* L. = DD) to produce hexaploid wheat, AABBDD (3-5). This happened about 8,000 years ago (1,2).

The origin of the B genome is presently unknown. From about 1958 through the 1960's, it was believed to have come from *T. speltoides* (Tausch) Gren. ex Richter (= *Ae. speltoides* Tausch), whose morphological characteristics were suitable and which grows in the proper area (6,7). However, *T. speltoides* has the wrong chromosomes (8,9) and the wrong cytoplasm (10).

Where did the B genome come from: There are at least four possibilities:

1. From a species that is now extinct.
2. From a species that has not yet been discovered.
3. From a diploid wheat; that is, the B is a modified A genome (11). However, the chromosomes of Johnson's assumed B-

---

*Cooperative investigations of USDA and the Missouri Agricultural Experiment Station. Paper No. 7679 of the Station Journal Series.

193

genome donor pair with those of the A genome, not the B (12, 13).

4. From two or more species. This is the view currently held by most students of the wheat group. By means of introgression the B genome of the original tetraploid wheat could have become greatly modified. Such introgression would most likely have occurred through hybridization of two or more tetraploids that both had the A genome but that had different second genomes (14). The common or "pivotal" A genome would have assured fertility of the hybrids, enabling segregated to arise having mixtures of chromosomes or chromosome segments from the different second genomes.

Although no breakthrough seems imminent concerning the origin of the B genome, there is ground for optimism that the problem will eventually be solved. One promising angle is to identify the donor of the B through the characteristics of its cytoplasm. It seems clear that the donor of the B genome also provided the cytoplasm of tetraploid wheat, for the tetraploids are male sterile when when put in the cytoplasm of diploid wheat. If cytoplasms are as slow to change as most people seem to think, then determination of which diploid species has the some cytoplasm as tetraploid wheat may reveal the B-genome donor. At least, this approach should eliminate all but a few of the potential donors. Several methods are now available for deciding whether two cytoplasms are identical. The traditional way, determining whether they interact in the same way with various sets of chromosomes, is still fruitful. A promising new way is to compare the fraction-1 protein of the respective chloroplasts (15). Another, probably more discriminating technique is to compare extracted, digested DNA of mitochondria (16) or chloroplasts.

Meanwhile, cytogenetic methods can presumably show the nature and degree of homology of B-genome chromosomes with those of various diploid species. The chromosomes of almost any diploid relative can be added, one at a time, to hexaploid wheat and studied for pairing with their B-genome homoeologues. It may be desirable in some cases to force pairing (by deleting or suppressing Ph, the inhibitor of homoeologous pairing) and to recover recombined chromosomes for analysis.

Whether or not the donor of the B genome is ever identified, the relatives of wheat are of potentially great importance to the future of wheat. It is to these species, mostly wild inhabitants of western Asia and the Mediterranean area, that we must increasingly turn for new genes for resistance to diseases and insects, for tolerance of drought and poor soil, and eventually even for increased productivity. Making these interspecific and intergeneric transfers of genes is not easy, but it can be done, and we can always hope to develop better methods. For example, a reliable technique for producing plants from microspores would be very useful.

Before expending much energy on exploitation of the relatives of wheat, most breeders will want to take advantage of
the considerable genetic variability still available in wheat
itself. Here they will be aided by a cytogenetic project
being conducted by the European Wheat Aneuploids Cooperative
(Coordinator: Dr. C. N. Law, Plant Breeding Institute,
Cambridge, England). This project has as its goal the cata-
loguing of all the genes in wheat that have quantitative
effects--cataloguing them not only as to their individual
effects and their interactions, but also as to where they are
located on the chromosomes. The successful completion of this
program could result in the conversion of wheat breeding from
an art to a science.

With the current concern for protein and amino acid
levels, it is interesting to note that deficiency for the
short arm of chromosome 2A results in about a 50% increase in
percentage of protein, with no apparent decrease in seed
weight (17). This of course suggests that high-protein mutations may be easily obtained in wheat.

What the future of Triticale and hybrid wheat may be is
difficult to predict. Triticale (the amphiploid of tetraploid
wheat and diploid rye) is reportedly grown on several hundred
thousand acres, and many knowledgeable people are optimistic
about its possibilities. Enthusiasm for hybrid wheat, on the
other hand, seems to have largely subsided. One of the problems with hybrid wheat is that full seed sets are almost never
obtained on the male-sterile lines, even under the best of
circumstances, and sets may fall to disastrously low levels if
weather conditions are unfavorable. This is surely due in
large part to the small anthers of wheat and the extremely
short lifetime of its pollen. Adapted as wheat is to self-
pollination, it is presumably uniform with respect to these
characters; but rye has very large anthers and long-lived
pollen. Why not, then, take full advantage of the rye component of Triticale and concentrate on hybrid Triticale rather
than hybrid wheat? It may be argued that since Triticale is
already a hybrid, no additional vigor can be expected from
crosses between Triticale cultivars. But wheat itself is a
hybrid in the same sense, and some wheat hybrids are exceedingly vigorous. Actually, since rye is self-sterile and
therefore an obligate out-pollinator, which shows pronounced
depression upon inbreeding, there is particular reason to
expect better performance when the rye chromosomes are heterozygous. Also, since it is now clear that vigorous lines can
be obtained that have one or more D-genome chromosome pairs
and only six or fewer pairs from rye, perhaps combinations
can be found that have mostly the wheat D genome and only
those rye chromosomes that are required for large anthers and

long-lived pollen.

I.   REFERENCES

1. Helbaek, H., *Econ. Bot.* 21, 350 (1966).
2. Helbaek, H., *Mem. Mus. Anthrop. Univ. Michigan* 1, 383 (1969).
3. McFadden, E. S., and Sears, E. R., *Rec. Genet. Soc. Amer.* 13, 26 (1944).
4. McFadden, E. S., and Sears, E. R., *J. Hered.* 37, 81 and 107 (1946).
5. Kihara, H., *Agri. and Hort.* 19, 889 (1944).
6. Sarkar, P., and Stebbins, G. L., *Amer. J. Bot.* 43, 297 (1956).
7. Riley, R., Unrau, J., and Chapman, V., *J. Hered.* 49, 91 (1958).
8. Kimber, G., and Athwal, R. S., *Proc. Nat. Acad. Sci. USA* 69, 912 (1972).
9. Gill, B. S., and Kimber, G., *Proc. Nat. Acad. Sci. USA* 71, 4086 (1974).
10. Maan, S. S., and Lucken, K. A., *J. Hered.* 62, 149 (1971).
11. Johnson, B. L., *Can. J. Genet. Cytol.* 17, 21 (1975).
12. Dvořák, J., *Can. J. Genet. Cytol.* 18, 371 (1976).
13. Chapman, V., Miller, T. E., and Riley, R., *Genet. Res.* 27, (1976).
14. Zohary, D., and Feldman, M., *Evol.* 16, 44 (1962).
15. Chen, K., Gray, J. C., and Wildman, S. G., *Science* 190, 1304 (1975).
16. Levings, C. S., and Pring, D. R., *Science* 193, 158 (1976).
17. Bozzini, A., and Giacomelli, M., *Genetics* 74, s29 (1973).

CURRENT THOUGHTS ON ORIGINS, PRESENT STATUS
AND FUTURE OF SOYBEANS*

T. Hymowitz and C. A. Newell

Crop Evolution Laboratory
Department of Agronomy
University of Illinois, Urbana

I. CURRENT THOUGHTS ON ORIGINS AND THE INTRODUCTION OF THE
    SOYBEAN INTO THE UNITED STATES.

    Historical and geographical evidence point to the east-
ern half of North China as the area where the soybean first
emerged as a domesticate around the 11th century B.C. Later,
Northeast China became a secondary germplasm center. Today,
the United States is the crop production center with the
State of Illinois as the leading producer of soybeans (1).
    Most probably the earliest introduction of soybeans into
the United States was the soybean grown in the Jardin des
Plantes, Paris. Paillieux (2) who traced the early attempts
to introduce the soybean into France revealed that in 1739
Compte du Buffon (George Louis Le Clerc) the director of the
Jardin des Plantes received soybean seed from missionaries
in China. The seeds were sown in the botanical garden the
following year and certainly so in 1779. Benjamin Franklin
who was the American Ambassador to France from 1778 to 1785
befriended Compte du Buffon and arranged for seed exchanges
between the two countries. The documented evidence for seed
exchanges between France and the United States is quite good.

---

*Contribution from Crop Evolution Laboratory, Department
of Agronomy, University of Illinois, Urbana, Illinois 61801.
Research supported in part by the Illinois Agricultural Experiment Station and the United States Agency for International Development under Contract No. AID/CM/ta-c-73-19.

In 1787, Buffon wrote a letter to Franklin in Philadelphia
acknowledging the seeds and rare plants sent to him from the
United States (3). In 1780, a letter written to Franklin's
grandson William Temple Franklin from de Malesherbes in Paris
mentions the shipment of three packages of seeds of plants
unknown in the United States (3). The introduced seed from
France was shipped to Philadelphia and planted in the botanic
garden.

The first mention of the soybean in American literature
is by Dr. James Mease in 1804 who wrote, "The soybean bears
the climate of Pennsylvania very well (4). The bean ought
therefore to be cultivated." In addition, in his book published in 1810 Mease noted, "The editor has the satisfaction
to assure them that the bean, Deliches Soyae bears the climate of Pennsylvania well."(5). Dr. Mease was very interested in agriculture, especially the introduction of animals
and plants from abroad (6) and he served for many years as
the secretary and then vice-president of the Philadelphia
Society for Promoting Agriculture. The agricultural society
was established in Philadelphia in 1785 (7). Dr. Mease must
have had access to information about the adaptability of soybeans in Pennsylvania - the soybeans most probably sent from
France by Benjamin Franklin.

The first citation seen for the soybean in the Midwest
is by A. H. Ernst (1853) of Cincinnati, Ohio, who wrote,
"The Japan Pea, in which so much interest has been manifested
in this country for a year or two past, from its hardihood to
resist drought and frost, together with its enormous yield,
appears to be highly worthy of the attention of agricultur-
ists (8). This plant is stated to be of Japan origin, having
been brought to San Francisco about three years since and
thence into Illinois and Ohio."

In the latter part of the 19th century, soybeans were
grown in Illinois for hay or as a soil fertility restorative
crop. One of the first farmers to grow soybeans in Illinois
was Mr. William H. Stoddard of near Carlinville in Macoupin
County (9). According to his daughter Mary, her father obtained soybean seed from the East and lectured on the virtues
of the soybean at farmers institutes. However, the value of
the soybean as a crop for Illinois was not unanimously accepted. For example, Robertson (1899) wrote, "The Soja bean has
been but recently introduced from Japan into our Northern
States. It is not a success in the South. Like the Cowpea
it will grow on any soil, and is relished by stock both for
its seeds and vines. The hay is inferior to that of the Cowpea on account of its heavy woody stems and because the
leaves rapidly fall off. Like the Cowpea also, the roots are
small and not much of value as a fertilizer (10)."

According to Burlison (11) soybeans were first grown experimentally by the University of Illinois in 1896 and in the
next year a circular about soybeans was published by Davenport (12). In 1903, another circular on soybeans was published by the University of Illinois (13). Planting, cultivation and harvesting procedures for soybeans were recommended in the circular. In addition, the results of the first
soybean variety trial in Illinois were reported. Of the 8
varieties tested 'Medium Green' was the outstanding yielder
at 41.7 bushels per acre. The variety was introduced into
the United States from Japan in 1899 by Prof. W. P. Brooks
(14).
     Although the soybean was tested in virtually all of the
states by the end of the first decade of the 20th century,
only about 2000 acres were planted in the United States in
1909 (15). The phenominal rise to prominence of the soybean
from the 1920's to today where it is the third largest crop
in the United States; the research and extension activities
of three University of Illinois agronomists - Burlison,
Woodworth and Hacklemann; and the processing, marketing and
use of the two primary products of the soybean, oil and meal,
has been fully documented in numerous publications (16-19).

II. PRESENT STATUS OF THE SOYBEAN - BREEDING AND GERMPLASM
     RESOURCES

     In the past four decades, soybean breeders and geneti-
cists in the United States diligently have labored to develop
varieties with high yield, resistance to diseases and other
plant pests, good agronomic characteristics and high oil and
protein contents. Recent sharp increases in palm oil imports
from Southeast Asia has had a depressing effect on soybean
oil prices in the world markets. Producers of soybeans get
paid on a weight basis regardless of the chemical content of
seed. If the soybean is to remain competitive, the system of
payment to the producer must be altered to reflect the market
value of the oil and protein content of the bean (20). In
the future, the greatest expansion in the use of the soybean
will come from the conversion of soybean protein into foods
for human consumption.
     The soybean industry in the United States is based en-
tirely on soybean germplasm introduced primarily from the
Peoples Republic of China, Japan and Korea. The United States
does not have any indigenous soybeans. Unfortunately, the
germplasm needed for the improvement of the soybean is rapidly being destroyed. The dwindling of these genetic resources
is due to the impact of modern technology and the population

explosion. The habitats for the wild and weedy relatives of the soybean are being covered by roads, houses, schools, factories and airports. Traditional farmers seeking ways to raise their standard of living are rapidly converting over from planting low yielding "land races" to superior high yielding varieties developed by modern soybean plant breeders. Soybean "land races" originated in traditional farming areas in a number of ways. Individual farm families have grown certain soybeans for hundreds of years, and soybeans containing specific traits have been grown year after year for religious, ceremonial or medicinal value.

The United States collection of soybeans is maintained by the U.S. Department of Agriculture in Urbana, Illinois and Stoneville, Mississippi (21). Dr. R. L. Bernard is the curator of the northern collection and Dr. E. E. Hartwig is the curator of the southern collection. Both collections are divided into four parts and they are as follows:

1. *Varieties.* This collection contains representative samples of each variety released in the United States and Canada.

2. *Genetic Stocks.* This is a collection of soybean lines which have specific traits whose inheritance has been determined.

3. *Wild Species.* This collection contains accessions of wild species in the genus *Glycine* (See Table 1).

4. *Plant Introductions.* This is a collection of seed of "land races" or improved strains from Japan, Korea, China and other countries donated to the United States and collected abroad by United States agronomists.

Altogether, the United States soybean collection contains about 6100 entries. Unfortunately, the 6100 entries represent only about 40% of soybean seed introduced into the United States. In the 1920's and 1930's there was no formalized group to take care of the collection hence many introductions were lost. The soybean collection is quite small when compared to the sorghum and rice collections, which contain about 15,000 entries and the wheat collection, which has over 35,000 entries.

The last United States plant explorers who visited China for the express purpose of collecting soybeans were P. H. Dorsett and W. J. Morse. They made their soybean collections from 1929 to 1931. Since then World War II and political considerations have prevented United States plant breeders from obtaining soybean germplasm in any reasonable quantity from the Peoples Republic of China.

Considering the vital role that soybeans play in the economy of the United States and the Peoples Republic of China it would appear that a perfect opportunity exists for

scientists in both countries to materially benefit from an
exchange program. The United States critically needs soybean germplasm from the Peoples Republic of China and the
Chinese can gain much from the United States soybean breeding programs.

III. THE FUTURE - EXPLOITATION OF THE GENUS *GLYCINE*

The genus *Glycine* Willd. to which the cultivated soybean
belongs is composed of three subgenera, *Glycine* Willd., Soja
(Moench) F. J. Herm., and *Bracteata* Verdc. (Table 1).
The generic name *Glycine* was introduced by Linnaeus in 1737
in the first edition of his Genera Plantarum (22) and based
on *Apios* Boerhaave, the plant now recognized as *Apios americana* Medik., but named *Glycine apios* by Linnaeus. *Glycine*
is derived from the Greek glykys, meaning sweet, and probably refers to the sweetness of the leaves and edible tubers
found in *G. apios* L. (23, 24). In later years all of the
eight Linnaean *Glycine* species except *G. javanica* L. were
transferred to different genera, including *G. apios* L. which
became *Apios americana* Medik. Until recently *G. javanica* L.
remained as the type of the genus. The origin of the name
Glycine thus bears little relationship to the species which
it now encompasses, as none of the current *Glycine* species
exhibit sweetness in their plant parts. The cultivated
soybean appears in Linnaeus' Species Plantarum (22) under
Phaseolus max L. (p. 727) and *Dolichos soja* L. (p. 727). After much initial confusion concerning the correct nomenclature
of the soybean (25-29) the new combination of *Glycine max* (L.)
Merr. proposed by Merrill (30) became widely accepted. It has
also been suggested that *G. soja* Sieb. & Zucc. is the valid
name for the wild relative of the soybean, this taking prior-
ity over the commonly used but later published *G. ussuriensis*
Regel & Maack (31).
Hermann in 1962 considerably clarified the classification
of *Glycine* by reducing the number of species described since
the time of Linnaeus from 286 to 10, excluding those species
which rightfully belonged to other genera (32). Further revision was necessitated by the discovery that Linnaeus' description of *G. javanica* the designated type for the genus,
was based upon a Pueraria (33). To minimize confusion, the
name *Glycine* was conserved from a later author, Willdenow (34),
and *Glycine clandestina* Willd. became the type for the genus.
Thus the last Linnaean *Glycine* species was removed from the
genus and transferred to *Pueraria montana* (Lour.) Merr. All
populations previously understood to be *G. javanica* L. were
placed under a new species combination of *G. wightii* (R. Grah.

TABLE 1

Chromosome Number and Geographic Distribution of Species in the Genus Glycine

Species	Chromosome Number	Geographical Distribution
Subgenus GLYCINE Willd.		
1. *G. clandestina* Willd.	40	Australia; South Pacific Isls.
1a. Var. *sericea* Benth.	--	Australia
2. *G. falcata* Benth.	40	Australia
3. *G. latrobeana* (Meissn.) Benth.	--	Australia
4. *G. canescens* F. J. Herm.	40	Australia
5. *G. tabacina* (Labill.) Benth.	80	Australia; South China; Taiwan; South Pacific Islands
6. *G. tomentella* Hayata	40, 80	Australia; South China; Taiwan; Phillipines
Subgenus BRACTEATA Verdc.		
7. *G. wightii* subsp. *wightii* var. *wightii* (R. Grah, ex. wight and Arn.) Verdc.	22, 44	India; Ceylon; Malaya; Java
7a. Subsp. *wightii* var. longicauda (Schweinf.) Verdc.	22, 44?	Arabia; Ethiopia; Congo Repub. to So. & W. Africa; Angola
7b. Subsp. *petitiana* var. *petitiana* (A. Rich.) Verdc.	22, 44?	Kenya; Tanzania; Ethiopia
7c. Subsp. *petitiana* var. *mearnsii* (DeWild.) Verdc.	22, 44?	Kenya; Tanzania; Malawi; Zambia
7d. Subsp. *pseudojavanica* (Tabu.) Verdc.	22, 44?	E. Africa; W. Africa; Congo Re.
Subgenus SOJA (Moench) F. J. Herm.		
8. *G. soja* Sieb. and Zucc.	40	China; Taiwan; Japan; Korea; USSR
9. *G. max* (L.) Merr.	40	Cultigen

ex Wight & Arn.) Verdc. Subgeneric names were altered according to the International code of Botanical Nomenclature to
reflect the changes in type (33).

The cultivated soybean *G. max* and its wild relative *G.
soja* together make up the subgenus *Soja*. Results of early
studies on chromosome number and size (35) and crossing experiments led Karasawa (36) to conclude that the cultivated soybean might have been derived from the wild bean. Accumulated
evidence from cytogenetics and morphology (37-41) and banding
patterns of seed proteins suggests that *G. soja* and *G. max*
are conspecific (1, 42) and supports the hypothesis that *G.
soja* is the wild ancestor of domesticated soybean. There
appear to be few, if any, barriers to gene flow between the
two species. The major differences exhibited by the domesticate are those qualities to be expected in a cultivated plant,
such as increased seed size, erect growth habit, larger plant,
and reduced shattering of seed pods at maturity. *Glycine
gracilis* Skvortz. was introduced in 1927 (43) as a new species, morphologically intermediate between the wild and cultivated soybean. Hermann, however, in his revision of the genus
did not feel that it justified specific rank and included it
with *G. max* (42). Hymowitz (1) has suggested that the weedy
G. gracilis evolved as a result of crossing between *G. max* and
G. soja, since it is found wherever *G. max* and *G. soja* overlap
in their distribution. The whole complex of wild, weed and
cultivated soybeans is analagous to that found in many plants
of economic significance (44).

The subgenus *Bracteata* is composed of *Glycine wightii*, a
variable species widely distributed throughout Africa and
parts of Asia. It has been used to some extent as a forage
crop and may be developed further as a pasture legume for
semi-arid and moderately humid tropical regions. Chromosome
numbers in the group have been reported as 2n = 22 and 2n = 44
(37, 45-48) compared with 2n = 40 or 2n = 80 for the other
subgenera. Differences in basic chromosome number and size
have given rise to speculation concerning the inclusion of *G.
wightii* in the genus (48). The chemical compound canavanine
is unique to the subgenus *Bracteata* (J. A. Lackey, 1973, unpublished manuscript), and differences have been found in
levels of seed compounds such as oils and fatty acids (49, 50).
A more detailed analysis of the subgenus is required to ascer-
tain whether its affinities with *Soja* and *Glycine* justify inclusion in the genus *Glycine*.

Six species and one subspecies are currently recognized
in the subgenus *Glycine* (32). The distribution is essentially
Melanesian, ranging from south China and Taiwan to Tasmania,
and eastwards to Tonga in the Pacific (Fig. 1). Three species are restricted to Australia. Since Hermann's major revi-

Fig. 1. Map of Australia showing the distribution of species within the subgenus Glycine.

sion of the genus, little information has been gathered at
the species level within the subgenus *Glycine*. Chromosome
number is 2n = 40 in diploids, and 2n = 80 in tetraploids
(37, 45, 48, 51); *G. tomentella* alone has been reported to
include both diploids and tetraploids. A major consideration
in the classification of species within the subgenus is the
presence or absence of a small rachis subtending the terminal
leaflet of the trifoliate leaves (32, 52, 53). Leaves are
said to be digitately trifoliolate when all three leaflets
are inserted at a common point on the petiole, or pinnately
trifoliolate when the terminal leaflet is removed from the
laterals by a short rachis. This distinction separates
G. clandestina, *G. falcata* and *G. latrobeana* with digitate
leaves, from *G. tabacina*, *G. tomentella* and *G. canescens*
with pinnate leaves. Other characters useful in delimiting
species include growth habit, leaflet shape and pubescence

type.

Glycine clandestina is distributed throughout the range
of the subgenus. It is diploid, with long, slender, twining
stems and oblong-lanceolate to occasionally oval-elliptic
leaflets; a pubescent form, G. clandestina var. sericea, has
also been recognized (53). Glycine falcata is diploid, characterized by erect or decumbent stout stems, oblong-lanceolate leaflets, falcate pods and a dense, strigose pubescence.
On the basis of morphology and certain seed proteins, G.
falcata appears to be the most singularly defined member of
the subgenus (54, 56) and is readily recognizable from herbarium specimens. Glycine latrobeana is restricted in its
distribution to south Australia and Tasmania. It typically
possesses short, erect or decumbent stems and obovate or
suborbicular leaflets. The chromosome number of this species
has not been reported. Glycine tabacina is characterized by
creeping or trailing stems, obovate or oval leaflets on the
lower parts of the plant, becoming elliptic-lanceolate to
narrowly oblong-linear on the upper portions of the stem.
The species is extremely variable morphologically, approach-
ing G. clandestina at one end of the range and G. tomentella
at the other. Bentham (53) described G. tabacina var.
latifolia with markedly obtuse and villous leaflets, but this
distinction was not maintained by Hermann (32). The transition from broad leaflet shapes at the base of the plant to
linear leaflets in the upper leaves, although present to
some degree in the other species, is particularly extreme in
G. tabacina, and serves to distinguish it from G. tomentella
(32). Chromosome studies have shown G. tabacina to be
tetraploid with 2n = 80 (37, 45, 48). However, some diploid
Australian populations appear to be closely allied morphologically with the G. tabacina group. Further analysis may
indicate that in Australia at least, G. tabacina encompasses
both diploids and tetraploids. Glycine tomentella exhibits
a range of distribution similar to that of G. tabacina.
Hermann in 1962 transferred G. tomentosa of Bentham (53) to
G. tomentella Hayata, G. tomentosa being a later homonym and
therefore untenable under the International Code of Botanical
Nomenclature (32). Slight morphological differences between
G. tomentosa Benth. and G. tomentella include linear-oblong
or lanceolate leaflets in the former, as opposed to elliptic
or rounded leaflets in the latter (55). On the whole, however, the group is rather more uniform than G. tabacina and
appears to be predominantly tetraploid. Stems are trailing
or climbing, and plants typically possess a tomentose-
villous pubescence which is often tawny in color. Several
collections from Australia have been reported as G. tomentella
with a chromosome number of 2n = 40 (45, 48). These plants

follow a distinctive pattern, with scrambling growth habit,
elliptic to rhombic leaflet shape, an appressed, sericeous
pubescence, and characteristic short, stout, few-seeded pods.
The group seems not to conform to *G. tomentosa* as described
by Bentham (53), and while it may be closer to *G. tomentella*
than to any of the other species, more information is required before the taxonomic status of this morphologically unique
diploid form can be properly evaluated. *Glycine canescens*
was proposed by Hermann in 1962 as a new name, the previous
G. sericea (F. Muell. in Hook.) Benth. of Bentham being a
later homonym and therefore untenable as a specific epithet.
The sole collection of this species available in the United
States was determined to be diploid (54). Plants of this
species characteristically possess twining stems, elliptic-
linear to narrowly oblong-lanceolate leaflets, and a dense
silky-strigose pubescence. *Glycine canescesn* appears to be
restricted to Australia.

It is evident from Fig. 1 that the subgenus *Glycine*
exhibits a somewhat disjunct distribution. The lack of recorded collections from the Philippines and Indonesia may be
due to lack of botanizing in those areas, or perhaps the species have been collected but filed under different genera in
herbaria owing to the confused taxonomic history of *Glycine*.
Alternatively the subgenus may have had a widespread distribution at one time, but owing to disturbance by man and other
environmental factors only relict populations remain. The
populations comprising these species are extremely variable,
and may not always appear unequivocally as one species or
another. Members of all six species produce cleistogamous
flowers with greatly reduced corollas in ones or twos in the
leaf axils. This tendency, together with the naturally high
incidence of self-pollination in the genus as a whole, ren-
ders a large proportion of seed to be selfed. As a result
many localized inbreeding populations arise, their degree of
morphological distinctness depending upon the parental source.
A low level of outcrossing would serve to introduce recom-
binant forms, which could again lead to distinct populations.
This situation is further exacerbated by the ability of
plants to root at those stem nodes in contact with the soil,
thereby enabling vegetative as well as sexual reproduction.

Despite the fact that the genus *Glycine* has constantly
undergone revision since the time of its conception, the present arrangement of subgenera and species may still not represent the final outcome. *Glycine soja* and *G. max* appear to
be conspecific, but from the practical standpoint of plant
breeders and geneticists it may be helpful to maintain some
degree of taxonomic separation. Further analysis of the *G.
wightii* group may justify its exclusion from the genus

altogether, but this has yet to be undertaken. Little is
known of the species within the subgenus *Glycine*. Variability within each species may be somewhat greater than hither-
to recognized, perhaps necessitating small revisions as more
information becomes available. Crossing experiments so far
have met with limited success, both within the subgenus
Glycine and between *Glycine* and *Soja* (46). At present, germplasm collections of the wild species are limited in the
United States, and should be expanded if a representative
sample of the variability available in the wild is to be obtained. Thus all problems in the genus *Glycine* are by no
means solved. Further work is needed to shed light on affinities between species in the subgenus *Glycine*, and between
the subgenera *Glycine* and *Soja* with a view to locating and
evaluating possible sources of useful germplasm for future
soybean breeding programs.

IV. REFERENCES

1. Hymowitz, T. *Econ. Bot.* 24, 408 (1970).
2. Paillieux, M. A. *Bull.*, *de la Soc. d'Acclimatation* Ser.
     3., 414 and 538 (1880).
3. Hays, I. M. Calendar of the Papers of Benjamin Franklin
     in the Library of the Amer. Philosoph. Soc., Philadel-
     phia, 1908. See Vol. III p. 348, Vol IV p. 42.
4. Mease, J. in "The Domestic Encyclopedia" Vol. V, p. 12.
     (A. F. M. Willich, Ed.) Philadelphia, 1804.
5. Mease, J. "Archives of Useful Knowledge" Vol. I, p.
     219. Philadelphia, 1810.
6. Rasmussen, W. A. "Agriculture in the United States, A
     Documented History" 414 and 471. Random House, New
     York, 1975.
7. Mease, J. "The Picture of Philadelphia" p. 256. Re-
     print by Arno Press, NY, 1970.
8. Ernst, A. H. "The Japan Pea" Report of the Commission-
     er of Patents, Agriculture 224 (1853).
9. Stoddard, M. H. Letter to Dr. W. L. Burlison, Agronomy
     Department, University of Illinois, Urbana. *Univ. of
     Ill. Archives*, Box 18, 8/6/2, 1944.
10. Robertson, L. S. *Illinois Agriculturist* 3, 25 (1899).
11. Burlison, W. L. Letter to Mr. George A. Montgomery,
     Coppers Farmer, Topeka, Kansas. *Univ. of Ill. Archives*
     Box 18, 8/6/2, 1943.
12. Davenport, E. "The Cow Pea and the Soja Bean" *Univ. of
     Ill. Agric. Expt. Sta.* Circ. 5, 1897.
13. Dalbey, D. S. "The cowpea and soybean in Illinois"
     Ill. Agric. Expt. Sta. Circ. 69, 1903.

14. Brooks, W. P. *Mass. Agric. Exp. Sta. Bull.*, 7, 14
    (1890).
15. Burlison, W. L. Speech given on August 21, 1942.
    Univ. of Ill. Archives.
16. Caldwell, B. E. (Ed.) "Soybeans: Improvement, Produc-
    tion and Uses" Amer. Soc. of Agronomy, Monograph 16,
    Madison, Wisc., 1973.
17. Markley, K. S. (Ed.) "Soybeans and Soybean Products"
    Interscience, New York, 1950.
18. Smith, A. K. and Cuck, S. J. "Soybeans: Chemistry and
    Technology" AVI Publishing Co., Westport, Conn., 1972.
19. Wolf, W. J. and Cowan, J. C. "Soybeans as a Food Source"
    CRC Press, Cleveland, Ohio, 1971.
20. Bullock, J. B., Nichols, Jr., T. E. and Updaw, N.
    Pricing soybeans to reflect oil and protein content.
    North Carolina Agr. Exp. Sta, Econ. Res. Rept. No. 37,
    1976.
21. Hymowitz, T. *Soybean Genetics Newsletter* 3, 29 (1976).
22. Linnaeus, C. "Species Plantarum." Vol. II. Stockholm,
    1753.
23. Henderson, P. Henderson's Handbook of Plants. New York,
    1881.
24. Henderson, P. Henderson's Handbook of Plants and Gen-
    eral Horticulture. P. Henderson and Company, New York,
    1910.
25. Lawrence, G. H. M. *Science* 110, (1949).
26. Paclt, J. *Science* 109, (1949).
27. Piper, C. V. *J. Am. Soc. Agron.* 6, 75 (1914).
28. Piper, C. V. and Morse, W. J. "The Soybean" McGraw-
    Hill, New York, 1923.
29. Ricker, P. L. and Morse, W. J. *J. Am. Soc. Agron.*
    40, 190 (1948).
30. Merrill, E. D. An interpretation of Rumphius' Herbarium
    Amboinense. Bureau of Printing, Manila, 1917.
31. Verdcourt, B. *Kew Bull.* 24, 235 (1970).
32. Hermann, F. J. "A revision of the genus *Glycine* and its
    immediate allies" USDA Tech. Bull. 1268, 1 (1962).
33. Verdcourt, B. *Taxon* 15, 34 (1966).
34. Willdenow, K. L. "Species Plantarum." Ed. 3. 3(2)
    1053. Berlin, 1802.
35. Fukuda, Y. *Jap. J. Bot.* 6, 489 (1933).
36. Karasawa, K. *Jap. J. Bot.* 8, 113 (1936).
37. Lu, Y. C. *J. Agr. Forest. (Taiwan)* 15, 1 (1966).
38. Tang, W. T. and Chen, C. H. *J. Agric. Assoc. China N.S.*
    28, 17 (1959).
39. Tang, W. T. and Tai, G. *Bot. Bull. Acad. Sin.* 3, 39
    (1962).
40. Ting, C. L. *J. Am. Soc. Agron.* 38, 381 (1946).

41. Weber, C. R. *Iowa Agr. Exp. Sta. Res. Bull.* 374, 767 (1950).
42. Mies, D. W. and Hymowitz, T. *Bot. Gaz.* 134, 121 (1973).
43. Skvortzow, B. V. *Manchurian Res. Soc. Publ. Ser. A. Nat. Hist. Sec.* No. 22, 1 (1927).
44. de Wet, J. M. J., and Harlan, J. R. *Econ. Bot.* 29, 99 (1975).
45. Cheng. Y. W. *Agr. Res. (Taiwan)* 12, 1 (1963).
46. Hadley, H. H. and Hymowitz, T. in "Soybeans: Improvement, Production and Uses" (B. E. Caldwell, Ed.) Amer. Soc. of Agronomy, Monograph 16, Madison, Wisc., 1973.
47. Lu, Y. C. and Thsend, F. S. *J. Agr. Forest (Taiwan)* 14, 13 (1965).
48. Pritchard, A. J. and Wutoh, J. G. *Nature* 202, 322 (1964).
49. Hymowitz, T., Palmer, R. G. and Hadley, H. H. *Trop. Agric. (Trinidad)* 49, 245 (1972).
50. Hymowitz, T., Collins, F. I., Sedgwick, V. E. and Clark, R. W. *Trop. Agric. (Trinidad)* 47, 265 (1970).
51. Kasha, K. J. *Taxon* 16, 146 (1967).
52. Black, J. M. Flora of South Australia. Part II. Ed. 2. K. M. Stevenson, Government Printer, Adelaide, 1864.
53. Bentham, G. Flora Australiensis. 2, 242-245 (1864).
54. Newell, C. A. and Hymowitz, T. *Crop Sci.* 15, 879 (1975).
55. Chuang, C. and Huang, C. "The Leguminosae of Taiwan: for pasture and soil improvement" Animal Industry Series No. U. Taipei, Taiwan, 1965.
56. Hymowitz, T. and Newell, C. A. *Illinois Research* 17, 18 (1975).

THE ORIGIN OF CORN –
STUDIES OF THE LAST HUNDRED YEARS

Garrison Wilkes

Biology II Department
University of Massachusetts, Boston
Harbor Campus
Boston, Massachusetts

I. INTRODUCTION

Corn is the most important food plant of the Americas and the single largest crop harvest in the United States (1-5). For three hundred years following the conquest of Mexico corn was considered a monotypic plant species. Then in 1832 Schrader published a species description under the name *Euchlaena mexicana* for teosinte. The close botanical relationships of teosinte with maize was recognized by botanists almost immediately. Ascherson, writing a hundred years ago, stated that teosinte was the nearest relative of maize and was unquestionably of New World origin; therefore maize must also be of New World origin (6-8,54). Ascherson (7) reviewed the variation in *Euchlaena* and emphasized its intermediate position between *Zea* and *Tripsacum*. A year later Ascherson (8) stated: In effect *Euchlaena* is a *Zea* (because of similarity of growth, leaves and tassel) in which the female flower is similar to *Tripsacum*. The English botanist Sir Joseph Hooker summarized the nineteenth century knowledge of teosinte: "In a botanical point of view *Euchlaena* is a most interesting genus, from its being the nearest congener of maize whose American origin it thus supports" (9).

Today, a hundred years later, the close relationship of corn and teosinte is widely recognized and based on their interfertility the two are cogeneric and placed in the genus *Zea*. Teosinte, *Zea mexicana* (Schrader) O. Kuntze, hybridizes readily with maize *Zea mays* L., and the F_1 hybrid is both robust and fertile. Both grasses are unique among the family in that the staminate flowers, borne in tassels, are separated from the pistillate flowers borne at a lateral position on the

stem. For the casual observer the two species are so similar
in appearance that the most reliable character separating the
two, the pistillate fruit, a distichous spike in teosinte and
a polystichous structure in maize. The seeds of teosinte are
dispersed as rachis-segments from the disarticulating spike.
This ability to disperse seed, an ability maize does not possess, distinguishes teosinte as a wild plant and because of
its unique reproductive strategy, a separate species from
maize.

II. RACIAL DIVERSITY IN MAIZE

Both teosinte ($2n=20$) and maize ($2n-20$) are highly variable, outcrossing, wind-pollinated species. There are approximately 200 recognized races of maize, 32 of which occur in
Mexico and 13 in adjacent Guatemala, with seven of this number
common with Mexico. The natural distribution of teosinte
falls within these two countries and is limited to primarily
the cultural area of ancient Mexican and Mayan civilizations
referred to by anthropologists as Mesoamerican.

All the maize known today is found only in the cultivated condition but archaeological evidence from Tehuacan indicates that one of the wild forms giving rise to two extant
Mexican races, Chapalote and Nal-Tel, was present in Mexico,
7,000 years ago. Yet even among the wild maize there appears
to have been racial diversity and recently Mangelsdorf has
proposed for maize 6 lineages: Palomero Tologueño (Mexico),
Chullipi (Peru), Confite Morocho (Peru), Kculli (Peru),
Chapalote/Nal-Tel (Mexico), and Pira Naranja (Columbia).
Mangelsdorf further states that "carried to a logical conclusion, it postulates at least six different races of wild corn
from which all present day races have descended" (10). Since
the founders of these lineages are geographically dispersed
it implies more than one domestication of corn from the diversity of the wild form. Also three of these lineages evolved
outside the known distribution range of teosinte and this pos-
es a special problem which is covered later in the paper in
the discussion of the role of *Tripsacum* in the evolution of
maize.

The lineages proposed by Mangelsdorf conform very well
with the Vavilov Centers for the origin of maize proposed by
the Russian plant breeder and genetist N. I. Vavilov, who
first studied these primary centers of genetic diversity, and
in some cases, secondary centers where the crop had undergone
rapid evolution (11). Clearly Mexico and Peru stand out as
centers from which the racial diversity of corn has spread.
Of the 32 races found in Mexico, 7 have counterparts in

Guatemala, 6 in Columbia, 5 in Peru, and 2 in Brazil (12).
Yet 20 of the 32, nearly two-thirds, are endemic to Mexico.
For Peru much of the same pattern of endemism exists, with
30 of its 48 races occurring only within its borders (13).

III.   GEOGRAPHIC DIVERSITY IN MAIZE

In the years from 1920 to 1940 the Russian plant breeder
N. I. Vavilov made over 2800 collections of maize in the New
World (11). Based on this collection Mexico was recognized
as the single center of greatest diversity. Like the Vavilov
centers for other crops, Mexico was characterized by mountain-
ous regions along the tropics long populated by agricultural
peoples, but isolated by steep terrain, arid regions, or
other natural barriers. This same reasoning was expressed by
Wellhausen et al. in The Race of Maize in Mexico (14, 15) to
account for the racial diversity found in Mexico: (1) the
preservation of primitive races, (2) the influx of exotic
races from countries to the south of Mexico, (3) hybridization with teosinte and (4) the geography of Mexico with its
varied habitats and isolating factors conductive to rapid
evolution.

Essentially the same conditions prevail in Peru but for
Vavilov the fact that teosinte was native to Mexico and not
found in Peru, favored Mexico for the origin. Vavilov considered teosinte to be the progenitor of maize and he attached
considerable significance to the fact that teosinte was fully
fertile with maize and that the naturally occurring hybrids
between the two could be found in Mexico.

Based on evidence very different from that considered by
Vavilov, Harshberger (16) came to the conclusion that maize
originated in Mexico and more specifically that it had once
been a wild plant in Central Mexico at elevations above 4,500
feet, in a semi-arid region with rains during the summer growing season of approximately 15 inches. The conclusions of
Harshberger's are remarkable because unknown to him they exactly circumscribe those areas of Mexico where the close relatives of maize are found and also they pinpoint the sites from
which our archaeological evidence of wild and early corn have
come.

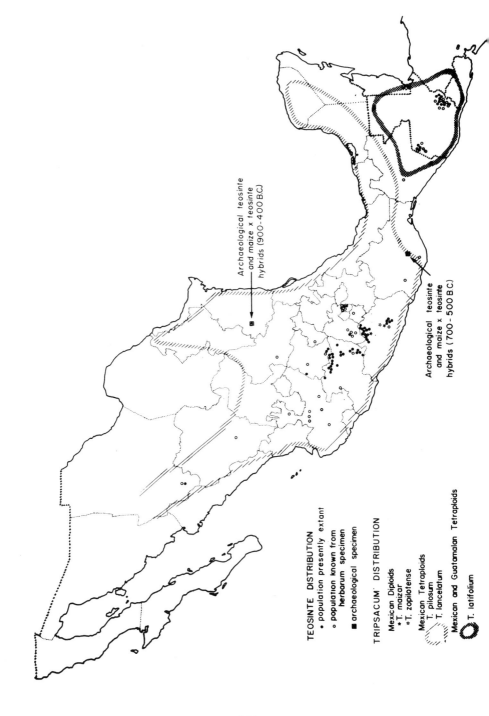

Archaeological teosinte
and maize x teosinte
hybrids (900-400 B.C.)

Archaeological teosinte
and maize x teosinte
hybrids (700-500 B.C.)

TEOSINTE DISTRIBUTION
• population presently extant
○ population known from
   herbarium specimen
■ archaeological specimen

TRIPSACUM DISTRIBUTION

Mexican Diploids
*T. maizar
*T. zopilotense

Mexican Tetraploids
T. pilosum
T. lancelatum

Mexican and Guatemalan Tetraploids
T. latifolium

214

IV. THE WILD RELATIVE OF MAIZE

Teosinte is a wild sometimes weedy naturally occurring
plant in Mexico and Guatemala. There are six recognized races
of teosinte, four of which occur in Mexico (Nobogame, Central
Plateau, Chalco, and Balsas) and two in Guatemala (Huehuetenango and Guatemala). In the Mexican states of Jalisco, Guana-
juato and Michoacan, teosinte is found mostly along stone
fences, bordering maize fields - not because it has invaded
the maize fields as a weed but because it is making a last
stand on this narrow strip of untilled soil. In a few local-
ities, such as Chalco in the Estado de Mexico, it has sucessfully invaded the maize field proper; but at several sites on
the Central Plateau where teosinte used to grow, a fact documented by specimens collected in the last century, it has
died out in recent times because of intensive land use in
farming, roadbuilding and pasturage.
The largest teosinte population and the one least likely
to disappear in the near future is that which occupies hundreds of square miles in the mountains of the Rio Balsas. The
teosinte of this region, along the western escarpment, (at
elevations of 800 to 900 meters), is the least maize-like of
all the teosintes found in Mexico; only the teosinte from
southern Guatemala is less maize-like than that of the Balsas.
The most maize-like of all the teosintes is Chalco, the maize-
mimetic teosinte of the Valley of Mexico.
In Guatemala there are two populations: Jutiapa (race:
Guatemala) in Southern Guatemala, which like *Tripsacum* tillers
profusely at the base and is distinctive among all the races
of teosinte in lacking a central spike in the tassel, and
Huehuetenango, in the north, which is more closely related
to Mexican races of teosinte than to Jutiapa.
The genus *Tripsacum* has assumed increasing importance
for research into the origin of maize ever since the hybridization of maize with *Tripsacum* was first reported by Mangelsdorf and Reeves in 1931 (17). Eight of the ten recognized
species are native to Mexico and Guatemala, a ninth, *T.
floridanum* is native to the tip of Florida, and the tenth
T. australe (and quite certainly as yet undescribed additional
species) is native to South America. The center of variation
for these perennial grasses, which unlike teosinte are quite
distinct in appearance from maize, is the western escanpment
of Central Mexico. This is almost exactly parallel in distribution to the large Balsas population of teosinte. The
habitat preferences of *Tripsacum* spp. in Mexico are nearly
identical to those of teosinte, seasonally dry, summer rains,
an elevation of about 1200 meters and limestone soils.
Tripsacum is usually placed in the tribe Maydeae along

with *Zea* (maize and teosinte) and these two genera are the
only New World members of this group of grasses (18, 19).
Tripsacum does show certain morphological resemblances to members of the Tribe Andropogoneae, particularly to the genus
Manisuris. The only significant difference between *Tripsacum*
and *Manisuris* is that *Manisuris* has perfect flowers while
Tripsacum has both male and female spikelets which are borne
distinctly, but which, unlike those of maize and teosinte,
are on the same inflorescence.

Evolution by polyploidy has been the mode in the genus;
again unlike maize and teosinte, which have followed an introgressive hybridization mode at the diploid level. The
diploid forms of *Tripsacum* are all morphically distinct and
allopatric in their distribution. The polyploidy forms are
not always easily distinguishable on either a morphological
or a geographic basis and there is considerable evidence to
indicate that they hybridize readily in the field (20).

Genetic experiments have established that exchanges can
and do occur between maize and *Tripsacum* chromosomes, even
though all evidence indicates that the recent evolution within
the genus *Tripsacum*, at least in Mesoamerica, has been independent and distinct from that of maize. It was with *T.
dactyloides* (2n=72) that Mangelsdorf and Reeves first successfully hybridized maize with *Tripsacum* (21). Since then *T.
dactyloides* (2n=36) and *T. floridanum* (2n=36) have been hybridized with maize. Studies of the hybrids have indicated
that certain segments of *Tripsacum* chromosomes can be substituted for corresponding segments in maize chromosomes and the
plants remain both viable and fertile. Galinat has mapped
more than 50 homologous loci on the chromosomes of these two
genera (22). The accumulated information on maize-*Tripsacum*
hybrids and their derivatives indicates that the respective
genetic architecture of maize (2n=20) and *Tripsacum* (2n=36, 72),
while quite different, are more similar than their karyotypes
would suggest.

V. THE ORIGIN OF MAIZE

The modern corn plant is a very productive, genetically
complex, and highly heterozygous plant whose wild form is extinct. The maize plant has been so altered in its basic biology that it is found only under cultivation. Often we think
we can better understand something if we know from where it
came or in terms of a living system, who their ancestors were.
Therefore the interest in the origin of corn, the mystery
comes because after a full century of speculation and research
there still is not a consensus. Either the ancestor of corn

was its closest relative, teosinte, or it was wild corn. For some this wild corn shared a distant ancestor with the present day races of teosinte. Somewhere in the middle of these two thoughts is a third origin of corn - a hybrid origin with an unknown grass and teosinte. A line up of these three theories over the last one hundred years follows:

Teosinte ancestor	by hybridization
Ascherson (54)	Harshberger (32, 33)
Beadle (23-25)	Collins (34, 35)
Galinat (26)	Schieman (36)
Harlan, de Wet (27, 28, 31)	
Iltis (29)	
de Wet, Harlan (30)	

common ancestor	maize from maize
Montgomery (37)	Mangelsdorf (10)
Weatherwax (38-40)	
Randolph (41)	

In a way the history of the origin of corn has come full circle over the last one hundred years. For Ascherson (54) the origin of maize was the transformation of teosinte spikes by fusion to form the ear. Hackel (1890) followed Ascherson and considered the ear of maize a monstrous or tetratological development based on fusion. This idea was supported by Goebel (1910) and Worsdell (1916), but Weatherwax (38), Mangelsdorf (1945), Kiesselback (1949), and Bonnett (1953) (See Reference 42) have failed to find any evidence of fascination in the maize cob so that on morphological evidence the origin of corn shifted away from teosinte.

In the 1890's both Harshberger (16) and Bailey (43) grew F_1 hybrid seed which had come from Mexico and concluded that genetic selection from teosinte accounted for the origin of maize. Both later recognized that maize was one of the parents of the F_1 and therefore the problem of the origin remained unsolved. About this time Kellerman (44) pointed out the homology of the central spike of the tassel with the ear and later Montgomery (37) extended the observation to teosinte. He concluded that both corn and teosinte have a common origin with the ear of maize being the homology of the central tassel and the fascicle of spikes in teosinte being the homology of the lateral branches of the tassel. Attention was removed from teosinte entirely by Mangelsdorf and Reeves (17, 45-47) who proposed that teosinte was a product of maize x *Tripsacum* hybridization and therefore did not figure it in the origin of maize. This view was widely adhered to for

thirty years. Although teosinte is not now considered a recent hybrid of maize x *Tripsacum* and can be once again seriously considered as an ancestor, Mangelsdorf (10) still excludes teosinte from the origin of maize.

Today, the most controversial element in any theory of
the origin of maize is the role of teosinte; that teosinte is
the closest relative of maize is universally recognized, but
there is not agreement about the relationship of teosinte to
maize. For Beadle (23, 24), Galinat (26), de Wet and Harlan
(30, 48), and Iltis (29) teosinte is clearly the ancestor and
wild form of maize. For them the structures that differentiate the two [polystrichous vs districhous ears, paired vs
solitary spikelets, naked vs covered grain] have taken place
under domestication. For Mangelsdorf in his recent book (10)
teosinte is a mutant form of maize and therefore contrary to
the teosinte ancestor theory "corn was the ancestor not only
of cultivated corn but also of teosinte (10)."

These two explanations: teosinte the ancestor, wild maize
the ancestor, are diagrammatically opposed, but the evolutionary events and morphological changes that would have taken
place are amazingly identical. Galinat has emphasized disrup-
tive selection, Beadle, selection under human guidance, Iltis,
the rational taxonomic imperative, and for Mangelsdorf spatial
isolation, as the mechanisms that preserved the separate evolutionary histories of these two genetically compatible sympatric grasses.

The tripartite theory of Mangelsdorf and Reeves (45) was
the basis of twenty-five years of innovative research and
widely read by those interested in the origin of corn. Although in recent years it has been demonstrated that teosinte
is not of hybrid origin, the other elements of the tripartite
hypothesis still stand. In fact there is universal acceptance
that modern maize is the product of the introgression of primitive maize with teosinte, and possibly with *Tripsacum* in
addition. The only controversial element about the pod/pop
primitive corn theory is the question of whether the ancestor
was wild maize or ancestral wild teosinte, and even here there
appears to be more disagreement over what the ancestor is
called, than on the traits it possessed.

Now that the hybrid origin of teosinte has been laid a-
side, the most favored alternative is a revived interest in
teosinte as the ancestral form. Teosinte is the closest relative of maize; it hybridizes with maize, the chromosomes pair,
crossing over between the two is virtually of the same order,
and the hybrids are fully fertile as are the subsequent back-
cross generations. The traits which distinguish teosinte
from maize are few and have to do essentially with the ear,
which would have been the part of the plant of greatest interest to food gatherers, and so some researchers view corn as

domesticated teosinte. While Mangelsdorf (10) views teosinte
as a derivative from wild corn, spatially isolated from the
ancestor of corn until after the domestication.

Although the exact ancestor is still not unequivocally
known, many of the aspects that surround the origin are commonly agreed upon by those holding different theories for the
origin. The domestication of corn appears to have (1) taken
place in Mexico about 10,000 years ago, (2) the habitat was a
highland site above 1500m., seasonably dry with summer rains
of about 25 to 50 cm., distributed over a three month period,
(3) the soil preference was for limestone derived soils often
on hilly land where the individual seed tightly enclosed by a
lower glume either in the rachis segment (teosinte) or by the
tunicate locus (wild maize) was distributed by gravity down
slope.

Since it is doubtful that either theory can reasonably
be expected to be proven experimentally, it is likely that
the most convincing evidence will come from complete archaeological profiles spanning the domestication of corn.

VI. ARCHAEOLOGICAL EVIDENCE

To date the oldest and most complete archaeological sequence for maize is from Tehuacan in Central Mexico. The
Tehuacan sequence spans the evolution of maize from prehistoric wild or near wild corn of 7000+ years ago to recent times.
The earliest cobs (5000 B.C.) are characterized by the uniformity of size and a bisexual condition, with the pistillate
spikelets below and the staminate spikelets, usually found
only in the tassel, at the tip of the ear. The cobs have
relatively long protective glumes that would have enclosed or
partially enclosed each kernel. The fragile rachis dispersed
the seeds. These characteristics are all thought to be those
of a wild plant, and indeed, these are just the characteristics that maintain teosinte as a wild plant in Mexican corn
fields.

The remains of later cobs are all larger and more varied.
In all of the botanical characteristics except size, the
earliest maize (3500 B.C. to 2300 B.C.) are virtually identical to the earliest remains. The increase in size is attributed to the improved growing conditions brought about by cultivation and irrigation. Then begins a period of teosinte introgression in the sequence and explosive evolution as measured by indurated lower glumes and a stiffening and elongation of the rachis tissue of the central axis in the cob. The
direct evidence of teosinte and maize x teosinte F_1 hybrids,
extend from 700-500 B.C. at Mitla Oaxaca and from Romero's

Cave (900–400 B.C.) in western Tamaulipus (49). These dates
for the Oaxacan and Tamaulipas sites are comparable to the
period at Tehuacan where maize exhibiting teosinte introgression was most abundant in the profile of the remains.

The stratified sequence of maize cobs in other archeolo-
gical sites where teosinte itself has not been recovered show
a clear and unmistakable change in morphology that is identical to the change induced by controlled introgression of teosinte into maize (50). Although the date of appearance of
indurated maize-cob remains, varies from one site to another,
the sequence of morphological changes following introgression
is uniformly the same (51, 53). The most extreme cobs exhibit characteristics of F_1 maize x teosinte hybrids, including
districhous arrangement of the spikelets and free solitary
cupules at the tip of the cob. The sudden and pronounced appearance of characters associated with teosinte introgression
in the cobs of a given level have been shown to be followed
by increased variability and evidence of heterotic effect in
succeeding levels. The evidence of teosinte introgression
into maize in the archaeological remains is circumstantial at
all but Romano Cave in Tamaulipas and Mitla in Oaxaca, yet
since indurated cobs with hard curved lower glumes and a stiff
rachis tissue can be matched by cobs of experimentally induced
hybridization with teosinte, and since maize and teosinte are
known to hybridize readily under natural conditions today,
there is no reason to doubt that the origin of indurated characters in the past are any different than that of the present
day. Indeed teosinte introgression is recognized in two-
thirds of the races of maize growing today in Mexico (15).

VII.   THE NEXT HUNDRED YEARS

The study of the origin of corn has come a full circle.
A hundred years ago teosinte was recognized as the closest
relative of maize and the possible progenitor. The hybrid
origin theories, and the tripartite hypothesis took the focus
away from teosinte. Now, the teosinte origin is being seriously considered again. The best evidence to resolve the origin story will probably have to come from archaeological profiles spanning the domestication of corn, notably from the
Balsas region of Central Mexico. Certainly there is little
disagreement that introgressed teosinte genes have contributed to both the racial diversity and the heterotic nature of
maize, but teosinte is found only in Mexico and Guatemala.
Where has the heterotic germplasm for South America come?
Quite possibly *Tripsacum* has entered the maize evolution directly. Further archaeological profiles are needed before we

know fully what has happened in the domestication of corn in
South America. To date, the earliest archaeological material
has resembled Mexican types but then it later varies radically
and is characterized by a sequence that is distinctly South
American. Did heterotic germplasm come from *Tripsacum* or did
unique structures come from genetic isolation from the maize
x teosinte evolution in Central Mexico? Certainly the conditions reported for the maize race Chococeno, in western coastal Colombia (52), which is broadcast grown on newly cleared
land, supports the idea of *Tripsacum* hybridization, and, in
addition, the plant possesses many attributes and grows not
too differently from a maize x teosinte segregate. Since teosinte does not occur within a thousand miles and *Tripsacum* is
common it has been assumed that Chococeno is the product of
the hybridization of maize and *Tripsacum*. This is a major
assumption and considering the importance of Chococeno to the
tripsacoid coastal races of maize in Ecuador and Peru now in-
trogressing into the races of higher altitudes it is appalling
that this dynamic hybridization has not been studies further.
The story of the evolution of maize has been introgressive
hybridization and it is intriguing to think that possibly
Tripsacum has done for South America what teosinte has clearly
done in Mexico and Guatemala.

VIII. REFERENCES

1. Sprague, G. F. (Ed.) "Corn; its Origin, Evolution and
    Improvement" 2nd ed. Agronomy Soc. Amer. Madison,
    Wisconsin. 1976.
2. Brandolini, A. in "Genetic Resources in Plants. Their
    Exploration and Conservation." (Frankel, O. and
    E. Bennett, Eds.) IBP Handbook No. 11, 1970.
3. Brown, W. L. in "Crop Genetic Resources for Today and
    Tomorrow." (Frankel, O. and J. G. Hawks, Eds.) Cam-
    bridge University Press.
4. Kempton, J. H. "Maize - Our heritage from the Indian"
    Rept. Smithson. Inst., 385 (1937).
5. Goodman, M. in "Evolution of Crop Plants" (Simmonds,
    N. W. Ed.) Longmans, New York, 1976.
6. Ascherson, P. *Bot. Vereins Prov. Brandenburg* 17, 76
    (1875). (also) 21 (1880).
7. Ascherson, P. *Sitzuber Ges. Naturf. Berlin, Proc.*
    160 (1876).
8. Ascherson, P. *Bull. Linn. Soc. Paris, Proc.* 105 (1877).
9. Hooker, J. D., "Fodder, plants: Teosinte" *Royal Garden
    Kew Rept.* 13 (1878).
10. Mangelsdorf, P. C. "Corn, Its Origin, Evolution and
    Improvement" Harvard University Press, Cambridge, Mass.

1974.
11. Valivov, N. I. *Bull. Appl. Bot. Genet. Pl. - Breed.* 16, 1 (1926).
12. Grobman, A., Salhuana, A., Sevilla, R. (in collaboration with P. C. Mangelsdorf) "Races of Maize in Peru" Nat. Acad. Sci. Nat. Res. Coun. Publ. 915, 1961.
13. Bukasov, S., *Bull. Appl. Bot. Genet. and Plant Breed.* 47, 141 (1930).
14. Welhausen, E. J., Fuentes, O. A., Hernandez-Corae, A., (in collaboration with P. C. Mangelsdorf) Razas de Maiz en la America Central. Folleto Tecnico No. 31 Sec. de Agricultura y Ganaderia, Mexico, 1957.
15. Welhausen, E. J., Roberts, L. M., and Hernandez-Xolocotzi E. (in collaboration with P. C. Mangelsdorf) "Races of maize in Mexico." Bussey Inst. of Harvard Univ., Cambridge, Mass. (1952).
16. Harshberger, J., *Contr. Bot. Lab. Univ. Pa.* 1, 75 (1893).
17. Mangelsdorf, P. C., and Reeves, R. G. *J. Hered.* 22, 328 (1931).
18. Rao, B. G. S. and Galinat, W. C. *J. Hered.* 65, 335 (1974).
19. Cutler, H. C. and Anderson, E., *Ann. Mo. Bot. Gard.* 28, 249 (1941).
20. Randolph, L. F., *Brittonia* 22, 305 (1970).
21. Mangelsdorf, P. C. and Reeves, R. G. *Bot. Mus. Leaflets. Harvard Univ.* 18, 389 (1959).
22. Galinat, W. C. *Rvol.* 27, 644 (1974).
23. Beadle, G. W. *J. Hered.* 30, 245 (1939).
24. Beadle, G. W. "Corn-Gift of the Gods", Garden Talk, *Chicago Hort. Soc.* 12 (1972).
25. Beadle, G. W. *Field Mus. Nat. Hist. Bull.* 43, 2 (1972).
26. Galinat, W. C. *Ann. Rev. Genet.* 5, 447 (1971).
27. Harlan, J. R. and de Wet, J. M. J. *Taxon* 20, 509 (1971).
28. Harlan, J. R. and de Wet, J. M. J. *Evol.* 17, 497 (1963).
29. Iltis, H., *Phytologia* 23, 248 (1972).
30. de Wet, J. M. J., and Harlan, J. R. *Euphytica* 21, 271 (1972).
31. Harlan, J. R., de Wet, J. M. J., and Price, E. G. *Evol.* 27, 311 (1973).
32. Harshberger, J., *Garden and Forest* 9, 522 (1896).
33. Harshberger, J. W., *Bol. de la Soc. Agr. Mex.* 23, 263 (1899).
34. Collins, G. N., *J. Hered.* 5, 255 (1912).
35. Collins, G. N., *J. Wash. Acad. Sci.* 8, 42 (1918).
36. Scheimann, E., "Entstehung der Kulturpflanzen" Berlin, 1932.
37. Montgomery, E. G. *Popular Sci. Monthly* 68, 55 (1906).

38. Weatherwax, P. *Bull. Torrey Bot. Club* 45, 309 (1918).
39. Weatherwax, P. *Bull. Torrey Bot. Club* 46, 275 (1919).
40. Weatherwax, P. "Indian Corn in Old America" MacMillan Co., New York, 1954.
41. Randolph, L. F., in "Corn and Corn Improvement", (G. F. Sprague, Ed.) Academic Press, New York, 1955.
42. Wilkes, H. G. "Teosinte: The Closest Relative of Maize" Bussey Institution of Harvard Univ., Cambridge, Mass. 1966.
43. Bailey, L. H. *Cornell Univ. Agri. Expt. Sta. Bull.* 49, 333 (1892).
44. Kellerman, W. A., *Meehan's Monthly* 5, 44 (1895).
45. Mangelsdorf, P. C., and Reeves, R. G. *Texas Agr. Exp. Sta. Bull.* 574, 1 (1939).
46. Mangelsdorf, P. C., *Proc. Am. Phil. Soc.* 102, 454 (1958).
47. Reeves, R. G., and Mangelsdorf, P. C. *Bot. Mus. Leaflets Harvard Univ.* 18, 357 (1959).
48. de Wet, J. M. J., Harlan, J. R., and Grant, C. A. *Euphytica* 20, 255 (1971).
49. Wilkes, H. G. *Science* 177, 1071 (1972).
50. Mangelsdorf, P. C., *Euphytica* 10, 157 (1961).
51. Mangelsdorf, P. C., MacNeish, R. S., and Galinat, W. C. *Science* 143, 538 (1964).
52. Roberts, L. M., Grant, U. J., Ramirez, R., Hatheway, W. H., Smith, D. L. (in collaboration with P. C. Mangelsdorf) "The Races of Maize in Columbia." Nat'l Acad. Sci./Nat'l Res. Counc. Publication 510, 1957.
53. Flannery, K. and Schoenwelter, J. *Archaeology* 23 (1970).
54. Ascherson, P. *Bot. Vereins Prov. Brandenburg* 21, 133 (1880).

INDEX